Message from the President

This is the first in a series of technology foresight reports identifying the strategic issues and opportunities in different sectors for members of the materials community of users, suppliers, designers and researchers.

The content is the output of the Energy Working Party established by the Institute of Materials under the chairmanship of Dr Brian Eyre F Eng FIM and takes account of advice and comments received in response to the Draft Report issued for consultation in December 1994. I should like to thank all contributors but particularly the members of the working party for their time and dedication to this work.

The Report is intended to assist those in industry, government and university who set priorities for the allocation of resources. The Institute will be promoting awareness of, and response to, the recommendations, and is willing to act as facilitator for initiatives where the various parties involved request it.

I believe members of the Institute and others less directly concerned with this industrial sector will find the report interesting and informative and I commend it to them.

Sir Geoffrey Allen FRS F Eng

Foreword

This report on technology foresight in the power generation industry in the UK was compiled from information generated by a working party appointed by the Materials Strategy Commission of the Institute of Materials, with the assistance of Quo-Tec Limited, an independent consultancy. The members of the Energy Working Party are:

Chairman	Dr B L Eyre	AEA Technology
Project Officer	Mr R C McVickers	Institute of Materials
Members	Dr M Ansell	University of Bath
	Mr R Conroy	Parsons Turbine Generators Ltd
	Dr P Kirkwood	British Gas Research
	Dr D J Naylor	British Steel plc
	Dr J C Whitehead	British Coal CRE (Coal Research Establishment)

They were assisted by:

Dr G B Gibbs	Formerly Research Manager, National Power
Dr R Judge	AEA Technology
Dr A D Batte	British Gas plc
Dr N A Waterman	Quo-Tec Ltd

CONTENTS   **Page**

CONTENTS   Page

List of Figures

List of Tables

EXECUTIVE SUMMARY

Of central importance to the continued health of the UK economy is the need to meet electricity demand economically and in a way that meets the growing environmental pressures. Looking to the future this requires the provision of plant based on cleaner and more efficient technologies. Such a requirement is not specific to the UK and applies increasingly to both the developed and the developing world.

This report describes the results of a foresight analysis focused on the materials technology requirements for power generation plants over the next 20 years. Although the analysis relates to UK requirements, it does address the wider global requirements and refers particularly to the links between the home and overseas markets in providing opportunities for UK industry.

To place the results of the study in perspective, Chapters 2 and 3 provide a broad strategic perspective emphasising particularly the substantial increase in electricity demand worldwide, the extent of the economic and environmental pressures, and the consequential need for advanced technologies to meet these pressures.

The main body of the report addresses the materials technology requirements for fossil fuel and nuclear power generation covering both fuel supply and power plant requirements, as well as the special back end requirements of nuclear power. Consideration is also given to renewables, drawing on the analysis carried out by the Energy Technology Support Unit for the Department of Trade and Industry, but this sector is not given detailed attention.

The report highlights the central role that is likely to be played in the long term by coal and nuclear for power generation, on the grounds of both fuel availability and environmental impact. Coal faces some major technological challenges in both fuel conversion, where integrated coal gasification is likely to be important, and clean coal combustion technologies. There are some key material technology requirements in all of these areas.

For nuclear power, there are unlikely to be radical changes in reactor design and the main material challenges here relate to long term reliability of key components. A major challenge is demonstrating the technology to meet the so-called back end requirements and particularly the treatment and disposal of radwaste, where material technology requirements relate to waste containment and validating repository performance.

The increase in the use of gas for power generation is likely to continue on both economic and environmental grounds. In the medium to long term, issues arise concerning the continued supply and costs of this fuel. Technology requirements centre on the need to extract gas from more remote and difficult locations and its transport to the point of use. There is also an incentive to improve the efficiency of combined cycle gas turbine plants, and to develop fuel cells for direct gas to electricity conversion.

The material technology requirements in all of these areas are highlighted in the report.

The final section of the report summarises our findings and conclusions. A key point highlighted is the decline in materials R&D to support power plant technology development in the UK in recent years. We consider this to be an issue of national strategic importance in terms of both meeting UK power generation requirements in the future and enabling UK industry to compete effectively in the growing international markets. We recommend that to address this issue there is a need to continue and extend partnerships between the key players - power generators, plant and materials suppliers and Government, not least because of its role as an important sponsor for R&D through, for example, the Research Councils. This collaborative approach, such as exists in other countries, is essential if UK industry is to compete in what will be a rapidly growing international market.

1. INTRODUCTION, METHODOLOGY, AND SCOPE

1.1 Introduction

The Materials Strategy Commission of The Institute of Materials has established a number of Working Parties (Energy, Aeronautics, Biomaterials) to examine materials technology requirements of UK industry as part of a 'foresight' exercise.

The Terms of Reference are the same for each of the Working Groups, viz:

- identify the major markets in their industrial sector for the short (1-4 years), medium (5-10 years) and long term;

and process technology and materials
y that will assist in addressing selected
es.

sults from the foresight analysis carried out by
and is concerned essentially with the large
sector, focusing in particular on the associated
rements.

ctricity has increased continuously over the
his demand is set to grow at an even faster rate
eveloping countries gather momentum. While
severe strains on resources and the
ovide opportunities for industrial companies
n plant who can meet the technological and
he future path taken in meeting domestic
n the role that UK fuel suppliers and plant
this global market. Thus, the Working Party
nship into account in the analysis reported

y was adopted:

rces influencing the Electricity Supply Industry
in the UK and elsewhere.

(ii)   Classify these influences under the following four main headings:

Economics: plant capital costs
                   operating costs
                   back-end costs

Environmental: emissions
                   pollution
                   waste management

Strategic: diversity
                   security of supply

Socio/political: macro-economics
                   liabilities
                   public acceptability

(iii)   Determine the likely consequences over short, medium and long term timescales in terms of demands for new and improved plant and equipment.

(iv)   Analyse the advances in materials technology required to meet the needs of the new and improved plant and equipment.

The sequence defined above recognises the hierarchy of influences which govern most technological developments, ie:

*   Market-led factors dominate the requirements for new plant and equipment.

*   The design of the new plant and equipment, including retrofitted components, dictates the needs for new and improved materials.

1.3 Scope

For the purposes of this study, the Power Generation Industry in the UK has been taken to encompass:

*   The industries concerned with the processing of fuels for large power generation plant, ie gas, oil, coal, and nuclear industries.

*   The Electricity Supply Industry itself.

- The manufacturers of plant and equipment for electricity generation and supply, ie manufacturers of boilers, turbines, burners, etc, and the materials manufacturers and processors which supply these industries.

The relationships between these industry sectors are illustrated schematically in Figures 1.1 and 1.2. The central role of materials R&D is shown in Figure 1.3.

1.4 Report Layout

The remainder of this report is divided into three sections:

Chapters 2 and 3 describe the strategic and structural factors influencing future electricity generation requirements, both globally and in the UK.

Chapters 4 through 8 describe the results of the foresight analysis of power generation and associated materials technology requirements from both the fuel supply and the plant supply viewpoints.

Finally, Chapters 9 and 10 consider the changing organisation of power plant materials R&D in the UK and what is needed to underpin UK power plant and materials manufacturers in meeting both national and global needs in the future.

Figure 1.1 Key Processes in the Life Cycle of Large Power Plant

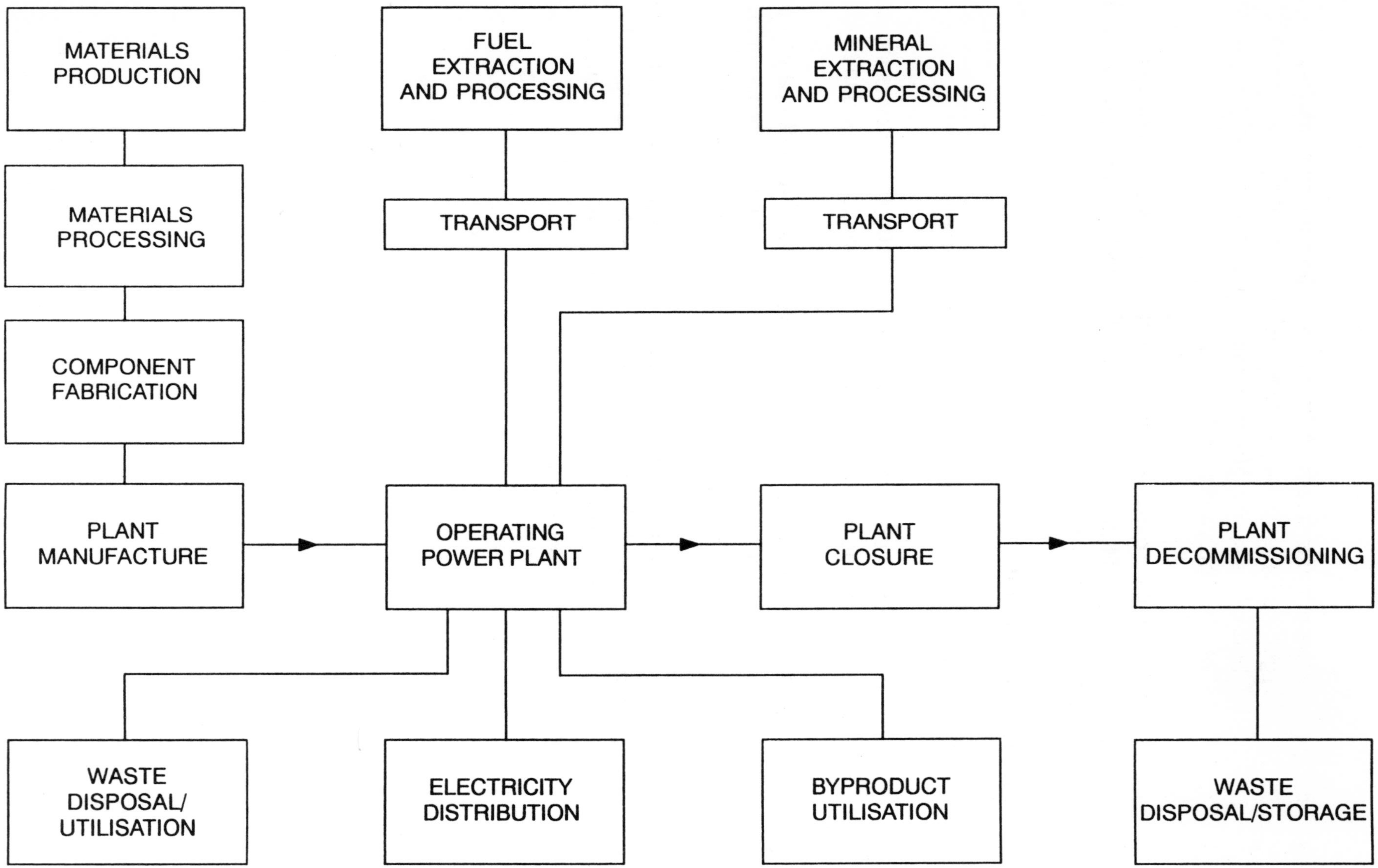

Figure 1.2   Electricity Supply in the UK; Principal Routes

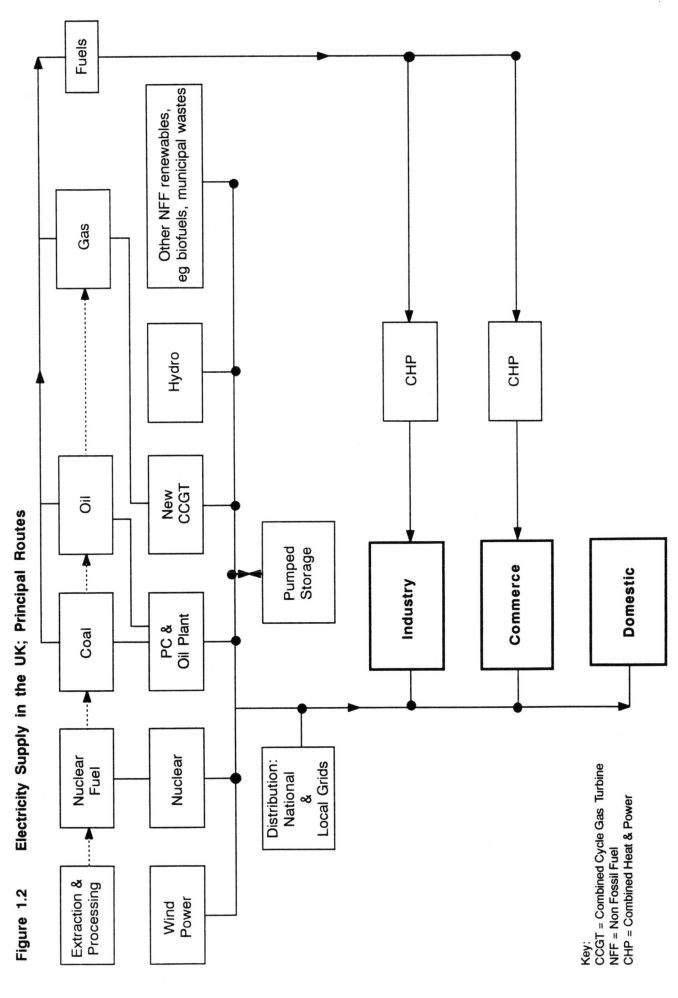

Key:
CCGT = Combined Cycle Gas Turbine
NFF = Non Fossil Fuel
CHP = Combined Heat & Power

Figure 1.3   Central Role of Materials R&D

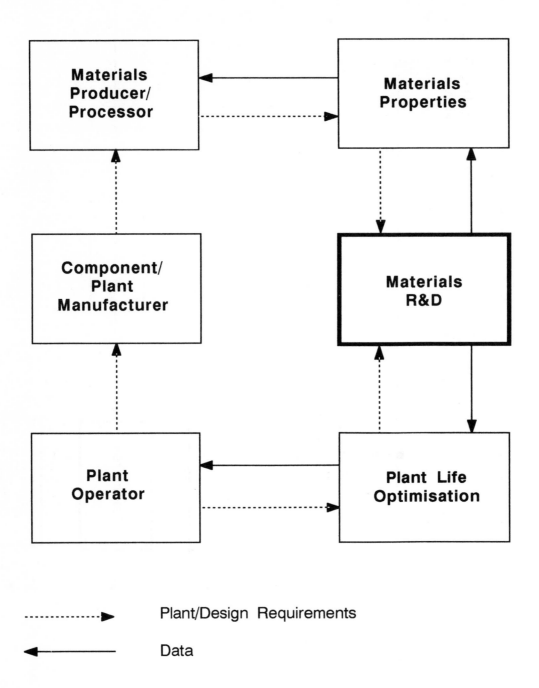

2. GLOBAL PERSPECTIVE

2.1 Energy for Tomorrow's World

The World Energy Council has recently published a widely acclaimed report of a detailed study of issues that will shape global energy supply and demand over the next several decades ('Energy for Tomorrow's World', WEC 1993). The report focuses on demand related issues against the background of major population growth, resource limitation, environmental pressures (particularly the enhancement of greenhouse gas levels through CO_2 emissions) and technology developments. The conclusions from the report are relevant in providing a broad strategic background to the present foresight study of the technology requirements of the power generation sector. The relevance for UK industry is in regard to:

a) the availability and price of fuels on the international market in the longer term,

b) overseas markets for UK plant and equipment manufacturers, and materials suppliers, and

c) environmental standards that new plant may be expected to meet.

2.2 Fuel Supply and Demand

The WEC report considered the ways in which global energy demand might be met in the future and compared these with current usage, using four different sets of assumptions. Three of the four cases are summarised in Table 2.1. Their 'reference case' B assumes modest growth in demand as populations expand, particularly in the developing countries. Case A has a higher rate of economic growth. Both A and B assume a greater rate of improvement in efficiency of energy use in developing countries than has been evident to date. If this assumption is removed, the 'modified reference case' B1 (not shown) is much closer to case A and almost identical to the Inter-Governmental Panel on Climate Change (IPCC) 'business as usual' scenario. Case C has modest growth, but a major global energy efficiency drive. Only case C keeps CO_2 emissions close to 1990 levels; for case A, they double by 2020, and increase by a factor of 1.5 for Case B.

Table 2.2 shows the estimated proven reserves in 1990, estimates of the reserves to production (R/P) ratios, and of the ultimately recoverable reserves. On the basis of such data, the WEC anticipates a continuing heavy reliance on coal, oil and natural gas for the next few decades.

Table 2.1

Energy source	1990, Gtoe	2020, Gtoe		
		A	B	C
Coal	2.3	4.9	3.0	2.1
Oil	2.8	4.6	3.8	2.9
Natural gas	1.7	3.6	3.0	2.5
Nuclear	0.4	1.0	0.8	0.7
Large hydro	0.5	1.0	0.9	0.7
'Old' renewable	0.9	1.3	1.3	1.1
'New' renewable	0.2	0.8	0.8	1.3
Total	8.8	17.2	13.4	11.3

'Old' renewable = traditional biomass (wood, crop residues, dung).
'New' renewable = solar, wind, geothermal, modern biomass, ocean, small hydro.

Table 2.2

Fuel	Proven reserves, 1990 Gtoe	R/P years	Ultimately recoverable Gtoe
Coal	496	197	3400
Lignite	110	293	
Conventional oil	137	40	200
Heavy crude			75
Bitumen			70
Oil shale			450
Natural gas	108	56	220

Note: (1) R/P = reserves/production ratio.
    (2) Table based on Tables 3.1 and 3.2 of WEC (1993).

However, the data indicate that even for the estimates of the ultimately recoverable reserves, the R/P ratios for oil and gas decrease for the higher energy demands projected for 2020 in Table 2.1, and that only coal is likely to be available in substantial quantities by the middle of the next century. Thus, anticipation of the decline in oil and gas reserves may raise energy prices 'well into the period between now and 2020'.

A further important factor impacting on energy planning is the geographical distribution of fuel reserves. Thus, it is significant that on the basis of 1990 estimates, only 4% of proven natural gas reserves are in West Europe, while more than 70% are in Central and Eastern Europe (CEE), the Commonwealth of Independent States (CIS) and the Middle East, which also contains about 70% of the currently proven oil reserves. Coal is more evenly distributed with China and India having huge coal resources (as well as potentially high energy needs) and a number of other developing countries have large coal resources.

The general conclusion to be drawn for the UK, and many other industrialised countries, is that with demand pressures and possible threats to imported fuel supplies resulting in future price increases, it is vital to maintain a strategic mix of primary fuels, particularly for electricity generation, and an adequate capability for exploiting indigenous fuels.

2.3 Electricity Generation

Turning more specifically to electricity generation, global demand has grown rapidly, both absolutely and as a proportion of total energy demand, doubling from around 15% in 1960 to 30% in 1990. Table 2.3 shows the consumption of commercial energy in 1988 for industrial countries and eight developing countries. The developing countries are Brazil, China, India, Indonesia, Malaysia, Thailand, Philippines and Pakistan. Looking to the future, it is anticipated that those countries will account for more than half the projected population growth of 3 billion and a significant fraction of global energy demand growth over the next 30 years.

**Table 2.3 - Consumption of Commercial Energy in 1988(%)
[from Fig 1.10 of WEC (1993)]**

	Industrial Countries	Developing Countries
Household and Services	21	21
Transport	22	14
Industry	19	34
Electricity	39	31

Fossil Fuels

As with other energy forms, electricity will depend substantially on fossil fuels and particularly coal for the foreseeable future. Worldwide, coal currently accounts for 40% of electricity generation, but the use of gas is increasing sharply, particularly in industrialised countries, with some 25GWe of new capacity currently being installed each year. Oil-fired generation tends to be favoured only in countries with cheap indigenous supplies, or where there is a shortage of alternative generating capacity.

The existence of substantial coal reserves in developing countries and the large reserves-to-production ratio for coal emphasise the need for new coal-fired generating capacity that is both efficient and clean.

Nuclear Power

Nuclear power generation currently accounts for around 16% of world demand. However, public confidence in nuclear power has been shaken by incidents at Three Mile Island and, more seriously, at Chernobyl, by reports of questionable safety standards in some countries of Eastern Europe, and by the debate concerning the true cost of the resource. Nevertheless, the WEC report speaks of the need for 'the early rehabilitation of nuclear energy' on a global scale. It goes on to assume that a significant fraction of growing world energy demand will be met by new nuclear plant, as indicated in Table 2.1. The alternative is even greater use of fossil fuels, with associated environmental and resource limitation problems.

Known uranium fuel reserves are estimated to match only 64 years of current world requirements at moderate (< US$130/kg) recovery costs, but this is greatly extended at higher recovery costs, and even more so by the use of reprocessing and recycling and the use of fast breeder reactors. When these are combined with estimates of the ultimately recoverable reserves, the total resource becomes almost twice that for all fossil fuels in Table 2.2.

Nuclear power use is expected to extend from the industrialised nations to the developing countries. The requirements for new plant, and the need to replace older plant that will be decommissioned in the next two decades, will provide major plant manufacturing opportunities.

In the UK, successive governments have considered nuclear power to be an important part of the strategically desirable diversity of supply, and one which also contributes to stabilisation of emissions of greenhouse gases. But, particulary in the context of the current electricity market, this support is subject to nuclear power being competitive and continuing to meet high safety standards.

Renewable Energy Sources

The WEC considers that the contribution from renewable energy sources is unlikely to exceed 2.9 Gtoe by 2020, though it could be higher (3.2 Gtoe) with strong government support.

In addition to limitations stemming from high capital cost, their report makes the important point that sources are often intermittent (eg wind power, solar power), leading to a requirement for:

a)   'overcapacity', and

b)   energy storage, especially where there is no back-up grid-
        connected system.

Nevertheless, renewables are potentially able to play an important part in meeting electricity demand under favourable conditions and it is important that the technology continues to be developed.

2.4 Environmental Impacts of Electricity Generation

It is well established that electricity generation has a major environmental impact, ranging from gaseous emissions from fossil fuel burning to back-end radwaste disposal from nuclear power, and various environmental consequences from renewables. Considering first gaseous emissions from fossil fuels, progress has been made in achieving cleaner power through the development of combustion technologies that minimise the production of SO_2 and NO_x and back-end technologies, SCR and FGD, that remove these combustion products from the flue gases.

A particularly intractable problem is how to deal with CO_2 emissions. Accumulated CO_2 levels are likely to increase globally for any realistic scenario for energy supply over the next 30 years or so, with consequent enhancement of the greenhouse effect and an uncertain degree of global warming. Electricity generation is important in this context because of its position as a major growing component of energy supply. The best prospects for containment of the CO_2 problem, at least in the medium term, will continue to come from greater efficiency in both energy supply and use. Removal of CO_2, although technically possible is, on the basis of present technology, very costly and ultimate disposal is problematical. The switch to gas alleviates the problem because of the low carbon content of the fuel, but it does not eliminate it. The only solution presently available on a substantial scale is the increased use of nuclear power, but there are currently other barriers to achieving this.

The predominant environmental issue for nuclear power is the management and ultimate disposal of radwaste. Although major advances have been made in the technology to achieve this, and technical solutions are available, there remain considerable political and social barriers to their implementation.

Governmental Responses to Environmental Issues

In their response to environmental concern, Governments face a number of short and longer term factors, including the need for competitive energy supplies, availability of different fuels, state of the economy and public acceptability of different options.

Within Europe, the UK is a signatory to the Large Combustion Plant Directive (LCPD). This requires members of the European Community to make progressive defined reductions in their total emissions of sulphur dioxide and nitrogen oxides. Major new plant is required to meet tight emission standards. In general these various restrictions will be met by some (country specific) combination of:

(i)   substitution of gas burning (in combined cycle gas turbine (CCGT) stations) for a substantial fraction of earlier coal burn;

(ii)   importing low-sulphur fuels;

(iii)   retrofitting flue-gas desulphurisation (FGD) equipment to major power plant (as at Drax and Ratcliffe in the UK);

(iv)   use of low NO_x combustion techniques, sometimes complemented with post-combustion NO_x removal (SCR);

(v)   the introduction of new advanced fossil plants meeting very tight emissions standards;

(vi)   non-fossil, especially nuclear, generation.

In addition, the regulatory authorities of an individual country may impose specific emission limits in the operating licence for any installation.

Regarding CO_2, many countries, including the UK, are signatories to the Framework Convention on Climate Change which requires measures aimed at reducing emissions to their 1990 levels by the year 2000. The precise way that this will be achieved is not clear. Certainly, there will be increasing pressure for energy conservation, increased process efficiency, and the use of technologies that minimise carbon dioxide emissions. It is also worth noting that incineration of waste may reduce greenhouse impact because decay in landfill sites releases methane which is a more potent greenhouse gas than CO_2. Also **sustainable** use of biomass can be neutral as regards **net** CO_2 emissions.

2.5   Market Implications

The WEC analysis of population growth distribution and associated energy demand reinforces the identification of a global market for new generating plant, particularly in the developing countries with fast growing economies. Such a market is likely to be very competitive and the standards required in terms of performance and environmental impact will be high. As already indicated, plants will need to match local conditions in terms of fuel availability and demand patterns, but there will be a major requirement for advanced coal plants and nuclear power is also expected to have an increasing role.

Although demand growth is likely to be much less in developed countries, there will be a need to replace existing plants as they reach end of life. This in itself will provide a major market. For example, in the UK all of the currently operating nuclear stations (excepting Sizewell B) and most of the existing coal stations reach end of life by 2020. An important factor with relation to the credibility of UK power generation plant manufacturers in the growing overseas markets will be their strength in the home market and their ability to sustain a development programme to meet the demands for advanced plants.

Against this background there is a global requirement for:

- High efficiency generation, with SO_2 and NO_x emission control when fossil fuels are used.

- Use of natural gas for power generation, when available and cost-effective.
- Nuclear power, subject to national policy.

- Electricity generation from renewable sources when cost-effective.

- Combined heat and power (CHP) schemes, in circumstances where it is possible to make use of the waste heat rejected by conventional power plant.

In broad terms, the associated challenges for materials R&D arise from:

- the need to ensure safe and reliable operation of nuclear plant;

- the need for higher temperature gas turbines to further increase the efficiency of CCGT plant;

- the need to increase temperatures and pressures of steam plant, to improve the efficiency of fossil-fuelled stations;

- specific requirements of high temperature and gas clean-up components of advanced coal utilisation plant;

- a need to reduce cost and/or increase reliability of emission control plant for conventional power stations;

- a requirement for lower cost reliable technologies for utilising renewable sources of energy.

14

3. THE UK ELECTRICITY SUPPLY INDUSTRY (ESI)

3.1 Introduction

Chapter 2 highlighted the substantial growth in world demand for electricity, particularly in the developing countries. This presents major business opportunities for the power generation plant industry which has been a traditional area of strength for the UK. Although growth in demand in the UK is projected to be modest, there will be a significant requirement for new plant as existing stations reach the end of life. The course adopted by the generating companies will, of course, impact on UK plant manufacturers in terms of orders placed with them and the associated technology requirements. This, in turn, will impact on their competitive position, and hence ability, to take a substantial share of the world market. It is therefore appropriate to review recent developments in the UK ESI and the consequences for the suppliers of both fuel and new plant technology.

3.2 Structure and Recent Developments

Most of the electricity used in the UK is supplied from large fossil-fired or nuclear power plant and in England and Wales is transmitted initially via the National Grid. (Scotland and Northern Ireland have vertically integrated systems). However, there is a small, but increasing, contribution from local generation. This enables the efficiency advantage of combined heat and power (CHP) systems to be exploited. At some industrial sites, surplus process heat can be used for on-site power generation. In general, the waste heat from large remotely sited power stations is not used because the necessary infrastructure is considered too costly or impractical to develop.

A further small, but increasing, fraction of the electrical energy requirements of the UK is supplied from renewable resources.

Market Forces and Influences

The UK ESI has, in recent years, been subject to influences which have transformed the industry in terms of its mission and business strategy and consequent need for new and improved technologies. These influences may be categorised under two main headings:

(i)   legislative, in particular privatisation and deregulation of the use of natural gas for electricity generation;

(ii)   environmental pressures, in particular, more stringent regulations concerning the emission of flue gases.

Privatisation

Before privatisation, the generation and transmission of electricity in England and Wales[1] was the statutory responsibility of a single body, the Central Electricity Generating Board. Distribution and utilisation was overseen by a separate body - the Electricity Council. After privatisation, non-nuclear generation was divided between National Power, PowerGen and a growing number of independent suppliers. Nuclear generating capacity remains in public ownership, with two separate companies, Nuclear Electric in England and Wales, and Scottish Nuclear.   A current Nuclear Review by the UK Department of Trade and Industry is to determine whether further nuclear stations should be built and whether the nuclear industry should also be privatised.

Transmission is the responsibility of a separate body - the National Grid Company - and distribution to customers is in the hands of the Regional Electricity Companies (RECs), formerly Area Boards, although some generators will have contracts directly with major industrial consumers. The RECs may themselves own and operate generating plant.

The mission of all of these companies is to make a profit and the generating companies compete throughout the day to supply a significant fraction of their electricity to the National Grid Company. They also compete with other major energy suppliers using gas, oil and coal as primary fuels to supply directly the major industrial consumers.

Two important changes relating to fuel supply for power generation accompanied privatisation. Throughout the life of the CEGB, the use of gas for power generation was discouraged. H M Government had powers to restrict the use of gas for electricity generation under Section 14 of the Energy Act, introduced to comply with the EC Gas Burn Directive. Both EC and UK regulations have now been repealed. Secondly, CEGB was obliged to purchase coal from British Coal at prices above world market prices. This obligation was initially transferred to National Power and PowerGen, but phased out by 1993.

An important result of these changes has been the so-called 'dash for gas', with new entrants opting for combined cycle gas turbine plant which has lower capital and overall (at current fuel prices) cost than new coal-

---

[1] Developments in Scotland have followed similar lines.

16

fired plant, and lower emissions of CO_2 and acid gases. The major generators, National Power and PowerGen, have also entered the CCGT market in order to remain competitive and to meet environmental constraints. The latter point is illustrated in Table 3.1, taken from National Power's evidence to the Select Committee on Energy in June 1991.

The building of CCGT stations is resulting in the closure of less cost-effective coal-fired capacity by the major generators, and a declining demand for British coal. In addition, lower cost coal imports have increased. These factors have been important, on the one hand, in improving the competitive position of British Coal and, on the other, in precipitating the UK pit closure programme.

The emergence of new generating companies, together with a major improvement in the performance of nuclear power, has led to a smaller market share for National Power and PowerGen. Thus, they have sought to strengthen their commercial position by expanding their operations overseas, both in the management of large power plant construction projects and as equity holders in such projects.

Non-Fossil Fuel Obligation

In order to meet the back-end costs of nuclear power, particularly of the older nuclear stations, and to encourage the development of clean renewable power generation options for environmental and strategic reasons, the UK Government has introduced some specific constraints into the electricity supply market. These are a surcharge on the price of electricity generated from burning fossil fuels (the Fossil Fuel Levy) and the Non-Fossil Fuel Obligation (NFFO). The latter requires the RECs to purchase a defined fraction of electricity for distribution from non-fossil fuel sources at subsidised prices. The original NFFO orders are due to expire in 1998, though the subsidy to renewables may continue beyond that date.

Environmental Legislation

An operating licence for large generating plant must be obtained from Her Majesty's Inspectorate of Pollution, which may apply site-specific constraints on emissions, and in the case of nuclear power plant from the Inspectorate of Nuclear Installations.

As noted already, the UK has emission reduction targets, for SO_2 and NO_x, that must be met under the terms of the European Community's Large Combustion Plant Directive. This applies to the total emissions

Table 3.1

COMPARISON OF ENVIRONMENTAL PERFORMANCE OF COAL COMBUSTION TECHNOLOGIES AND GAS COMBINED CYCLE (CCGT)

	Design efficiency (% LHV)	Sulphur dioxide (g/kWh)	Nitrogen oxides (g/kWh)	Carbon dioxide (g/kWh)	Solid products	
CONVENTIONAL COAL						
with FGD	39.5	1.2	2.3	830	Ash	50
future with FGD	42	1.0	2.3	780	Gypsum	25
future with FGD + SCR*	42	1.0	0.3	780		
FLUID BED CYCLES						
simple	39.5	1.1	1.2	830	Reactive	
circulating combined	40	1.2	1.2	820	ash/gypsum/	
pressurised combined	42	1.0	1.1	780	lime mixture	90
future 'topping'	46.5	0.9	0.7	710		
COAL GASIFIER CYCLES						
integrated combined (IGCC)	43	0.1 - 0.5	0.8	760	Ash or slag	50
future IGCC	45	0.1 - 0.5	0.8	730	Sulphur	5
GAS COMBINED CYCLE (CCGT)	52	-	0.8	380	None	

Subcritical steam cycle conditions are assumed. All figures are approximate and depend on the exact plant specified.

* All NO_x emissions can be further reduced by the addition of selective catalytic reduction SCR.

from all plant exceeding 50 MW (thermal). The targets have been translated into total emission limits for the large generating companies. All new plant is individually subject to SO_2 and NO_x emission limits under the LCPD.

As a signatory to the Framework Convention on Climate Change, the UK is also committed to attempting to reduce CO_2 emissions to their 1990 levels by the year 2000. This is a much broader issue than use of fossil fuels for power generation. It is being addressed in all energy sectors primarily via encouragement for more efficient fuel utilisation, and for fuel switching where appropriate. In the power generation sector, encouragement is being provided by the NFFO and Fossil Fuel Levy, and via the deregulation of natural gas as a fuel.

3.3 Future Requirements for Power Plants

Figure 3.1 illustrates the effects of the various market forces on the plant and equipment needs of the UK ESI. The main technology requirements are expanded in the following sections.

Emission Control Plant

The major operators of fossil-fuelled power plant, National Power and PowerGen, have installed FGD plant at two large coal-fired power stations - Drax and Ratcliffe. However, their apparent intention is to meet the requirements for further emission reduction arising from the LCPD via an increased use of low-S imported coal and, especially, via construction and operation of combined-cycle gas turbine (CCGT) plant burning natural gas with concurrent closure of less cost-effective coal-fired plant. Further FGD retrofits in the UK therefore seem unlikely, unless required by HMIP to deal with a major operational change (eg introduction of a high-S fuel such as Orimulsion into an oil-fired station) or for some other local reason.

The NO_x emission limits of the LCPD can be met by the low-NO_x combustion modifications made to their large coal-fired boilers by National Power and PowerGen, without the need for retrofitting catalytic NO_x reduction technology. NO_x emissions from CCGTs also satisfy the current statutory requirements for new plant.

New Gas Turbine Plant

With current fuel prices, and a capital cost per MW for Combined Cycle Gas Turbines (CCGTs) being less than half that for other fossil fuelled

19

Figure 3.1   Map of Market Influences on Electricity Supply Industry for the UK

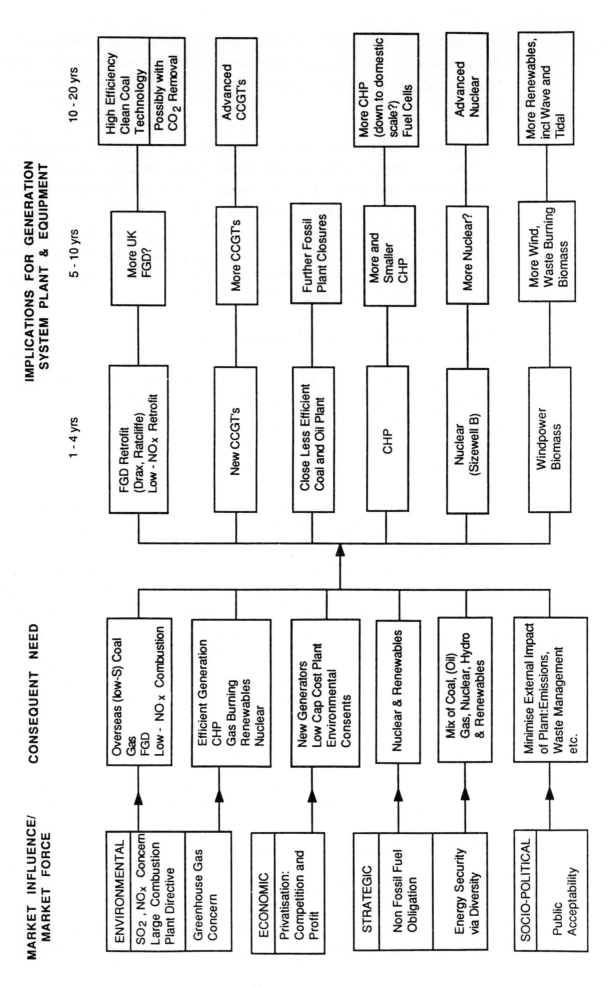

and nuclear plant, CCGTs are the natural choice for new plant. Environmental constraints are also more readily met. New entrants and the major private generating companies (National Power and PowerGen) formed from the CEGB, are now supplying electricity generated by CCGTs and additional plant is under construction. 11% of fossil fuelled electricity generation in the UK was supplied by CCGTs in 1993 compared with only 1.4% in 1992. This is set to increase very substantially to 40-50% of the fossil fuel generated electricity (ie around 33% of the total) by the late 1990s.

New Coal-Fired Plant

It seems unlikely that any major new coal-fired plant will be built in the UK in the next ten years, unless special funds are made available for demonstration of advanced technologies. Beyond that time, National Power and PowerGen will need to replace some of their ageing coal-fired stations that have remained operational. Accordingly, Figure 3.1 shows a box for 'High Efficiency Clean Coal Technology'. Generators' plans for coal-fired plant replacement are not in the public domain, but recent European assessments suggest two major options: an evolutionary development of current pulverised-coal plant with an advanced (ultra-supercritical) steam cycle, low-NO_x combustion, and FGD, or a step change to some form of advanced coal combustion plant, probably operating in combined cycle mode.

Despite the lack of an internal requirement, the major UK generators do have a current interest in clean efficient coal-fired generating plant - evolutionary development of existing plant to supercritical and ultra-supercritical steam conditions, on account of their involvement in overseas projects. Longer-term interests are presently served by, for example, National Power's (4%) stake in ELCOGAS - an IGCC Demonstration Plant in Spain.

Nuclear Power Plant

Nuclear power generation in the UK has traditionally relied on British designs of gas-cooled reactors, first the Magnox stations and then the Advanced Gas-Cooled Reactors (AGRs). However, only two new gas-cooled reactors (Heysham 2, Torness) have been built in the last 20 years and three of the older Magnox stations have now closed.

In recent years, the contribution of nuclear power to electricity supply has been less than suggested by installed capacity, owing to early operating problems with the new AGRs. However, performance improved dramatically between 1992 and 1993 (see Tables 3.2 and 3.3) and AGRs now rank alongside the world's best in terms of load factor. Nuclear power currently accounts for about a quarter of the electricity supplied to the grid.

In addition to the gas cooled reactors, one other new nuclear station has just been commissioned at Sizewell by Nuclear Electric. This is based on a Westinghouse design of pressurised water reactor (PWR), modified to meet the UK requirements, particularly on safety.

The generators' plans for the future use of nuclear power will be influenced by the outcome of the UK Government's Nuclear Review. It will need to be competitive and continue to meet high safety standards. Nuclear Electric has expressed full confidence in the PWR for any new plant to be built and, in partnership with Westinghouse, is seeking to promote this plant overseas. Perceived uncertainties, particularly concerning back-end costs (which the nuclear industry claims to be small) will have to be balanced against the environmental advantages of the reduction in acid gas and CO_2 emissions, and price stability due to the relative insensitivity of total costs to fuel price. As emphasised in Chapter 2, the WEC view is that nuclear power is likely to have an increasingly important role in electricity generation. Decisions reached with regard to our future domestic supply options will clearly influence the role that UK industry will be able to play in this world market.

Combined Heat and Power (CHP) Schemes

Combined Heat and Power (CHP) schemes achieve high efficiencies of energy utilisation, as high as 90%, because they make use of waste heat. The Energy Technology support Unit (ETSU) has claimed (Energy and Environment Paper No 3) that up to 50% of total UK electricity demand could be met by small-scale CHP schemes: packaged units based on gas-fired internal combustion engines (20 to 500 kW) and gas turbine plant (up to 10 MW). However, this was said to require 'further evolution of the technology' and a more realistic estimate was set at 10%.

The 1993 figure for installed CHP capacity in the UK was 2.9 GWe. UK Government considers a realistic target for the year 2000 to be 5.0 GWe.

Whilst there are obvious industrial and commercial markets for CHP, the possible domestic market is more problematical. Large housing developments do not have the necessary infrastructure to use the waste heat, nor is it being built into new developments. A very small unit suitable for individual home application does not yet exist, but such a development must be of interest to suppliers of natural gas.

Table 3.2   **Declared Net Capacity, MWe; all UK Generators**

	1991	1992	1993	1994
Conventional fossil	54,644	51,520	47,841	44,506
CCGTs	76	331	1,334	6,163
Nuclear	11,353	11,353	11,353	11,894
GTs	3,130	2,963	2,539	2,018
Hydro	1,400	1,412	1,425	1,433
Pumped storage	2,787	2,787	2,787	2,787
Other renewables	125	153	220	317

Table 3.3   **Plant Use as M tonnes Oil Equivalent Used**

	1991	1992	1993
Coal	52.61	46.90	39.53
Oil	7.53	8.43	6.10
Gas	0.63	1.73	7.61
Nuclear	17.43	18.45	21.49
Hydro	0.39	0.46	0.37
Other	1.81	1.11	1.24

Source: DTI Digest of UK Energy Statistics, 1994.

Other Non-Fossil Fuels

Renewable electricity generation presently accounts for only a small fraction of total UK supply. However, the fossil fuel levy linked to the NFFO has made it attractive commercially for new entrants. The NFFO is scheduled to be phased out, but by then other factors may encourage use of renewable sources, eg tighter environmental legislation, cost penalties on landfill for waste disposal, and falling plant and equipment costs.

As a consequence, the next ten years is likely to see new plant requirements for onshore windpower, waste-to-power schemes, biomass combustion or gasification, and use of landfill gas, as well as hydro-power. Beyond ten years, offshore windpower schemes and use of tidal power are both conceivable. In the medium term, this will not provide a major market for suppliers of large generating plant/components, but it will represent a significant market for some specialist plant manufacturers.

4. COAL

4.1 Introduction

Chapter 2 noted that from the viewpoint of security of supply, it is desirable that a mix of fuels is used for power generation in the UK, with indigenous coal reserves making a major contribution. If this is to be maintained in a free market economy, locally-produced coal must compete in price with imported coal and with alternative fuels. Additionally, new plant must be available to meet both environmental and cost constraints, when existing coal-fired capacity needs to be replaced.

Chapter 2 also discussed the likely growth in demand for electricity in the developing countries and the role that increased coal production will play in meeting that demand. Developments in the techniques for coal extraction and processing will therefore be of benefit on a global scale, and manufacturers of coal-fired power plant will have challenging opportunities in a world market.

4.2 Industry Structure and Market Influences

The UK coal industry was nationalised in 1947, at around the same time as the ESI, and is currently following the ESI into privatisation. The sale of all mining assets is scheduled for completion by the end of 1995.

In the 1980s, coal burning accounted for 80% of electrical power generation in the UK, and this in turn accounted for around 80% of the market for domestic coal. However, UK coal production has fallen from 110 Mt per annum in 1983 to 50 Mt per annum in 1994. As indicated in Chapter 3, this is a consequence of the 'dash for gas' and of the improved performance of nuclear stations.

For much of its history, UK power station coal has been more expensive than imported coal, but sales were guaranteed as a result of agreements between the nationalised industries. These agreements have gone, but recent improvements in the performance of the coal industry mean that it can now compete with internationally traded coal at inland UK power stations. In addition to closure/re-organisation of the least cost-effective operations, introduction of new technology has played an important role. Further advances will be necessary to maintain an ability to compete with overseas coal and other fuel sources.

The declining UK electricity market for coal has been a consequence of not only the cost of the fuel but also:

FIGURE 4.1   GENERAL MAP FOR COAL SUPPLY AND USE IN THE ENERGY FIELD

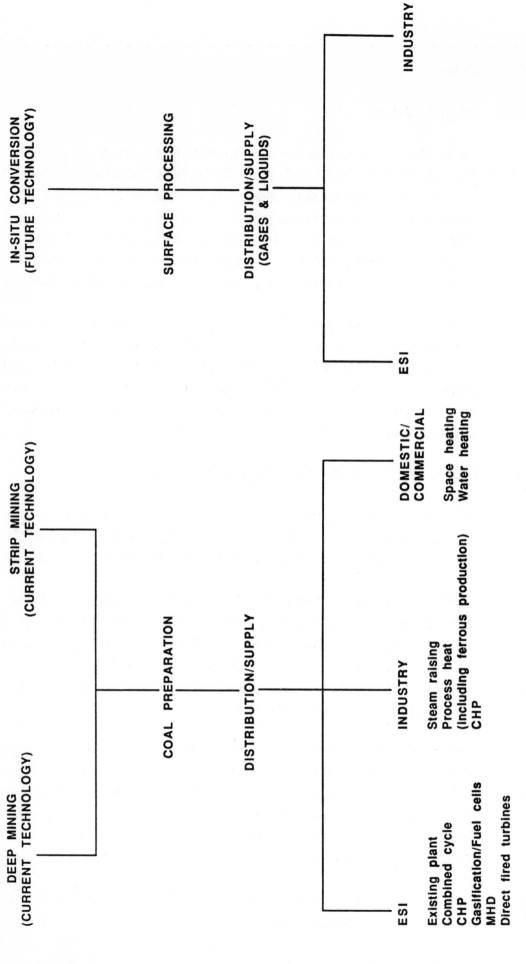

a)   the availability of relatively cheap supplies of natural gas, and the relatively low capital cost of CCGT plant;

b)   the higher capital cost of pulverised coal power plant, and the additional cost of (FGD or other) plant for removing SO_2 from flue gases.

Thus, there is a need for improvements in coal utilisation technologies in order to maintain or even increase coal's share of the ESI market.

4.3   Coal Production and Processing

Production

The structure of the coal production and supply industry is outlined in Figure 4.1. Approximately 80% of coal production in the UK is achieved by deep mining; the remainder being from strip mining. Looking overseas, strip mining, with its substantially lower costs, is the principal mining method in a number of countries, for example, Australia.

In-situ conversion of coal, for example to gaseous or liquid products, is only at the R&D stage. It provides a more socially acceptable method of coal extraction, although careful consideration will have to be given to its potential environmental impact.

The principal activities involved in the extraction of coal by deep mining, strip mining and in-situ conversion are summarised in Figure 4.2. Each of these activities, potentially at least, provides opportunities for new and improved technology and therefore has need of technology foresight. The impact of the various market forces on future developments is mapped through to materials R&D requirements in Figure 4.3.

Coal   Cleaning

One of the major environmental problems of coal burning arises from the fact that coals often have a substantial sulphur content (around 1.6 wt% for UK coals), leading to a requirement for costly flue gas desulphurisation plant.

Coal 'washing' is already used to reduce the sulphur and ash content of coals, but other physical and chemical processes are being developed that can give a cleaner fuel - albeit at increased cost.

27

FIGURE 4.2   MAP of COAL SUPPLY

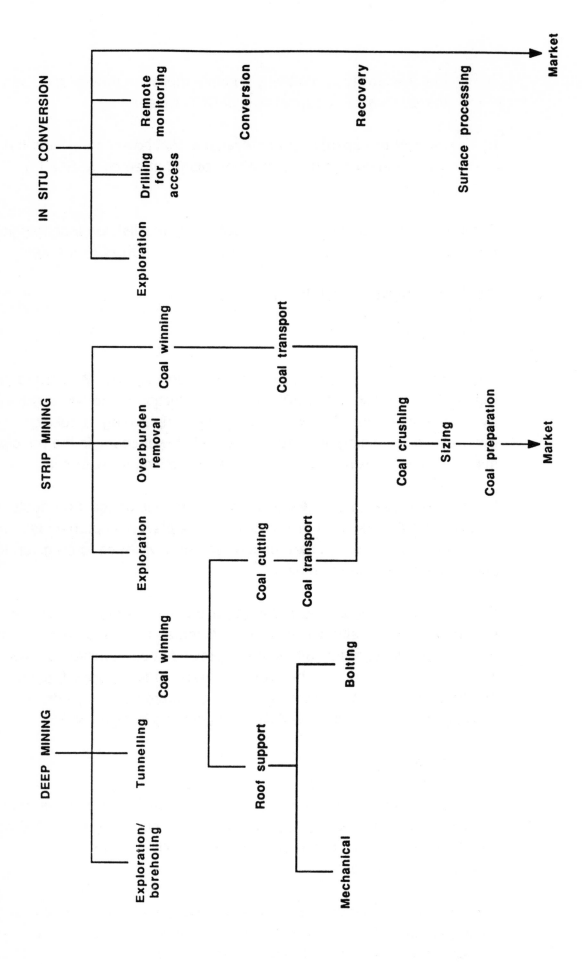

FIGURE 4.3 MAP OF MARKET INFLUENCES ON COAL SUPPLY INDUSTRY

	MARKET INFLUENCE (MI)	CONSEQUENT NEED (CN)	MATERIALS REQUIREMENT	
			SHORT/MEDIUM	LONG TERM
ECONOMIC	• Lower Cost	• Higher Drivage/Cutting Rates	Improved wear and fatigue characteristics	
		• Improved Reliability		
		• Reduced Maintenance		
	• Lower Demand	• Pit Closure		
ENVIRONMENTAL	• SO$_x$ Concern	• Low - Sulphur Coal (Open Cast)		
	• Tighter Planning Control	• Less Strip Mining		
		• Improved Coal Preparation		
STRATEGIC	• Energy Security	• Continued Long Term Coal Use		
	• Greater Use of Offshore Supplies	• In-Situ Conversion		Improved high temperature corrosion/erosion resistance
SOCIAL	• Opposition to Strip Mining	• More Deep Mined Coal		
	• Improved Safety	• Improved Materials Performance	Improved wear and fatigue characteristics	

An alternative approach is to gasify the coal and clean up the product to give a synthetic substitute for natural gas, usually termed syngas. At present, syngas is much more expensive than natural gas.

Gasification

The WEC report has drawn attention to the fact that world coal reserves greatly exceed those of oil and natural gas, and that by 2020, scarcity may have increased the price/reduced the supply of natural gas for power generation. At some point it will become economic, and eventually it will become essential, to generate synthetic gas from coal.

Gasification techniques are already well developed and a number of different gasifiers are commercially proven at large scale. They may be air or oxygen blown, fixed bed, fluidised bed, or entrained flow. Materials issues for gasifiers relate especially to the integrity of ceramic linings of pressure vessels and ductwork. As indicated below, currently available gas clean-up techniques require prior cooling of the gas. The additional cost of the syngas cooler increases the cost of the product gas, and hence postpones the time when a stand-alone gasifier would become economic.

Somewhat earlier introduction of coal gasification for large scale power generation may occur as a result of the development of integrated gasification combined cycle plant, where heat wasted in a simple gasification process is recovered in the combined cycle.

4.4 Coal-Fired Power Plant

Figure 4.4 maps market influences through to potential requirements for new plant. The materials development required to make these options viable is discussed below.

Advanced PC Generating Plant

Large-scale power generation from coal has traditionally used combustion of pulverised coal (pc) in a furnace to raise steam, at temperatures up to around 580°C, to drive a steam turbine. A subcritical steam cycle has been preferred in the UK, but in other countries new coal-fired plants have been supercritical plants for the last decade or more. Plant efficiency is reduced slightly if FGD is installed, but design efficiencies of 40 - 42% are now possible, even with subcritical steam conditions.

Figure 4.4

MAP OF MARKET INFLUENCES ON COAL USE IN THE ESI

	MARKET INFLUENCE (MI)	CONSEQUENT NEED (CN)	PLANT SUPPLY REQUIREMENT (PSR) SHORT/MEDIUM	LONG TERM
ECONOMIC	• Lower Cost • Lower Demand	• Cheaper Materials		• High Temperature Turbines • High Temperature Gas Clean Up
ENVIRONMENTAL	• LCPD • NO_x/SO_x Concerns	• Low - S Coal Use • Instal FGD • Other Sulphur Retention • Low NO_x Capability	• Supercritical Steam	
	• Greenhouse Gas Concern	• Increased Efficiency • Reduced CO_2 Emission	• More CHP • FGD Retrofits	• CO_2 Removal • Coal Fired Fuel Cells • MHD Development
STRATEGIC	• Energy Security • Availability of Supplies			
SOCIAL	• Government Influence eg Subsidies & Taxes			

Note: With reference to Fig 4.1, the ESI (rather than industrial or domestic use) is the area in which the greatest benefits will be achieved through the development of new materials.

PC plant seems likely to remain a competitive option for overseas projects in the short term, and for both overseas and the UK in the medium term. It is now possible to design ultra-supercritical (USC) plant with maximum steam conditions of 300 bar/580°C, giving design efficiencies of 45 - 46%. This has been made possible by the development of improved ferritic steels (T91 for superheaters, P91 for thick section components). It remains important to continue to obtain data on the performance of these steels.

Higher plant efficiencies are in principle possible in the medium or long term: 47% at 325 bar/620°C or even 50% at 325 bar and 650-700°C.

Some step changes are required to achieve these higher efficiencies. At the lower temperatures, low alloy ferritic steels are used for the water panels. At higher steam temperatures and pressures, greater creep and corrosion resistance is required. The method of construction implies that the material developed must be capable of being welded without pre-heat or post-weld heat treatment. More highly alloyed ferritic steels are being investigated/developed for this purpose (HCM12A, HCM2S), and this is important for the longer term.

The more advanced USC steam cycles will require austenitic superheaters with improved creep strength and corrosion resistance achieved by minor alloying additions. Again, various candidate steels exist, but with inadequate performance data.

For the same advanced plant, thick section boiler components may be constructed from 9 - 12% Cr steels with higher creep resistance than P91. Various steels are under test (NF 616, TB 12M, HCM 12A) that are suitable for plant with steam temperatures up to 600°C. A British Steel development E911, based on a high temperature variant of P91, is at an earlier stage of testing. Further development is required, possibly the new steel NF 12, for USC conditions of 325 bar/630°C.

High alloy (10% Cr) ferritic steels are available for steam turbine rotors and conditions up to 325 bar/610°C, though more performance data is required and is currently being obtained in the CEC's COST 501 programmes. New improved ferritic steels are being developed in this programme, and it may be possible to use these up to 650°C. The COST 501 group is also undertaking a preliminary study of materials suitable for plant operating at 700/720°C and 375 bar. It is very probable that nickel based and austenitic materials will have to be developed for many components.

Figure 4.5 Materials Development Requirements of Advanced Steam Cycles

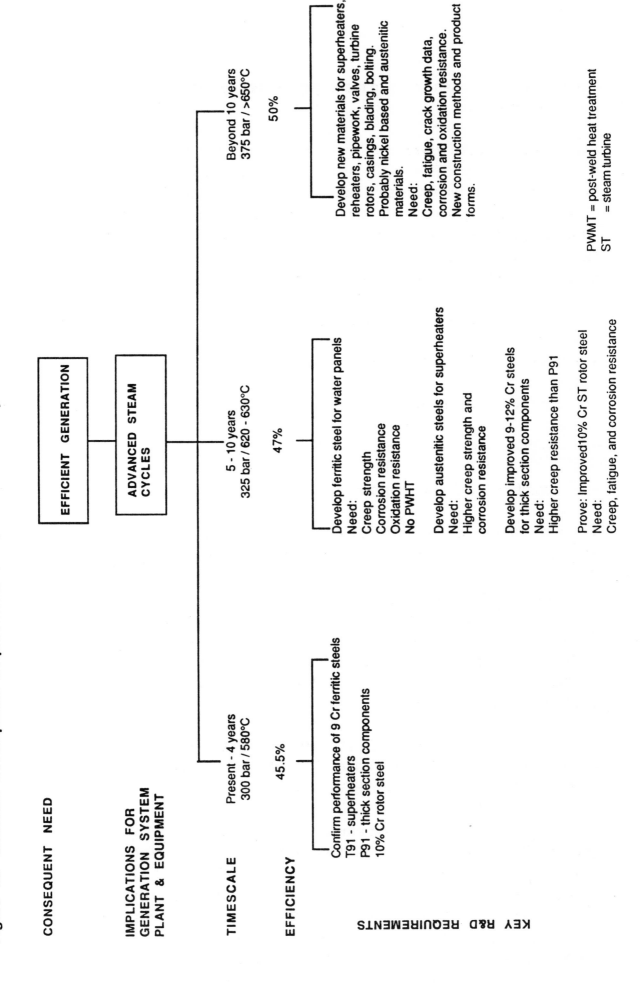

CONSEQUENT NEED

EFFICIENT GENERATION

IMPLICATIONS FOR
GENERATION SYSTEM
PLANT & EQUIPMENT

ADVANCED STEAM
CYCLES

TIMESCALE

Present - 4 years
300 bar / 580°C

5 - 10 years
325 bar / 620 - 630°C

Beyond 10 years
375 bar / >650°C

EFFICIENCY

45.5%

47%

50%

KEY R&D REQUIREMENTS

Confirm performance of 9 Cr ferritic steels
T91 - superheaters
P91 - thick section components
10% Cr rotor steel

Develop ferritic steel for water panels
Need:
Creep strength
Corrosion resistance
Oxidation resistance
No PWHT

Develop austenitic steels for superheaters
Need:
Higher creep strength and
corrosion resistance

Develop improved 9-12% Cr steels
for thick section components
Need:
Higher creep resistance than P91

Prove: Improved 10% Cr ST rotor steel
Need:
Creep, fatigue, and corrosion resistance

Develop new materials for superheaters,
reheaters, pipework, valves, turbine
rotors, casings, blading, bolting.
Probably nickel based and austenitic
materials.
Need:
Creep, fatigue, crack growth data,
corrosion and oxidation resistance.
New construction methods and product
forms.

PWMT = post-weld heat treatment
ST   = steam turbine

33

Figure 4.5 summarises some of the key materials development requirements.

FGD and de-NOx Plant

In order to comply with emission regulations, new advanced pulverised coal plant may require low-NO_x combustion plus flue gas treatment to remove sulphur dioxide and, in some countries, to remove additional NO_x, as well as standard particulate removal by electrostatic precipitator or baghouse filter. Options for sulphur dioxide removal are:

• sorbent injection into the furnace

• wet scrubbing of flue gas

• spray dry scrubbers

• regenerable removal processes

• combined SO_2/NO_x removal.

The first four are commercially 'proven', the final one is at the demonstration stage. The most common sorbent is limestone, and the most common by-products are calcium sulphite/calcium sulphate/ash mixtures, gypsum (hydrated calcium sulphate) and sulphuric acid (from a regenerable process). It is generally believed that wet scrubbing will dominate FGD process selection for new large power plant in the short and medium term, with gypsum being the major by-product. In the UK, limestone-gypsum systems were chosen for both Drax and Ratcliffe power station retrofits.

An interesting challenge for materials technology is to find new large-scale uses for gypsum, since production would readily exceed current demand with major expansion of FGD use.

Operating conditions in FGD systems are onerous, flue gas atmospheres being highly erosive and corrosive, even more so if fuels are high in chlorine as is the case for UK coals. Plant operators require the minimum lifetime costs commensurate with high availability of plant, and this hinges on materials technology issues. Ducts, absorber vessels and heat exchanger surfaces may be protected by choosing a suitable alloy or by surface coating / use of liners. Depending on position in the circuit, protection may be given by use of rubber liners (T-restrictions), glass flake reinforced plastics (may spall under thermo-mechanical stressing), vitreous enamelled steels (for reheater systems), or Ni-based alloys. Chimneys may need to be relined because of the higher moisture content of gas post-FGD. Proving least cost solutions will continue to be an important area of materials R&D.

There is considerable overseas experience, in Germany and Japan particularly, with de-NO_x systems that rely on selective catalytic removal (SCR) of NO_x, by reaction with ammonia over a catalyst bed. Here, a materials technology challenge is to produce lower cost catalysts that retain their mechanical integrity and have adequate working life. Additional developments of high surface area sorbents and catalysts are required for combined SO_2/NO_x removal systems.

The above requirements are summarised in Figure 4.6.

Fluidised Bed Combustion

In the simplest type of power station using fluidised bed combustion of coal, the fuel is injected into a fluidised bed operating at atmospheric pressure. Heat is removed from the combustion gases via a bank of steam tubes and used to drive a conventional steam turbine. With a 'bubbling' bed, the tube bank is placed in the combustor. In a circulating bed, the heat exchanger is separated from the main combustor, placing less onerous requirements on erosion and corrosion resistance. Sulphur dioxide emissions can be reduced by adding limestone to the bed, and NO_x emissions reduced by combustion control.

Fluidised beds can be designed for efficient combustion of poor quality coals, other fuels and wastes. Around the world they have been favoured for small (< 100 MW_e) units, when the fuel supply was cheap and FGD costs were to be avoided. Circulating atmospheric beds are now developed to utility scale (100+ MW_e). However, they do not offer efficiency advantages over conventional pc plant, and they compare unfavourably with new gas-fired CCGT plant in the UK. Materials considerations arise, especially in regard to corrosion and erosion resistance of heat exchanger surfaces.

FBC plant efficiency can be increased to around 42% by pressurising the combustor and operating the plant in combined cycle mode, driving a gas turbine with filtered hot gases from the combustor. There is scope here for improving the erosion resistance of turbine blades and/or improving high temperature ceramic filters. Though various pressurised circulating bed designs exist, it is the bubbling bed PFBC pioneered at Grimethorpe in the UK that is now being offered commercially (by the Swedish

Figure 4.6   Map Relating Consequent Need & Materials Technology

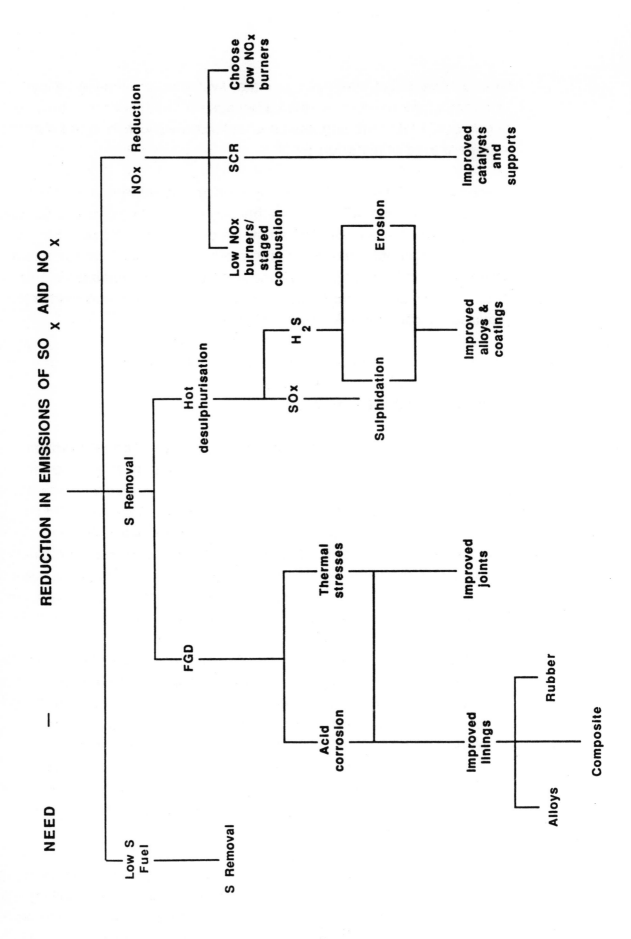

company, ABB Carbon).

Whilst these types of plant may capture specific markets in the short and medium term, in the long term they may be superseded by gasification or partial gasification options - see below.

First Generation IGCC

First generation integrated gasification combined cycle plant is at the utility demonstration stage, eg the 300 MWe Demkolec plant using a Shell gasifier at Buggenum in the Netherlands. Design efficiencies of 43-44% are possible. These types of plant require heat exchangers to cool the raw gas produced by the chosen gasification process so that it can then be subjected to low-temperature gas clean-up processes. To avoid efficiency losses, the heat is used to raise steam which is integrated into the steam cycle of the combined cycle plant. Corrosion resistance requirements of the heat exchangers are particularly onerous, since the gas will generally contain entrained ash, and chlorine and sulphur compounds. This makes ferritic boiler steels unsuitable for the higher temperature sections (above 300°C). Either corrosion resistant coatings must be used, or austenitic steels with a high chromium content, plus other alloying additions for improved corrosion resistance. However, there is a lack of long term data on performance, particularly at superheater temperatures.

As far as the UK is concerned, by the time that replacement coal-fired generating plant is required, first generation IGCC may be unable to complete with USC developments of pc plant, owing to its high capital cost. Competition could arise from second generation IGCC or other advanced plant options, however.

Second Generation IGCC

On a 10 year timescale, alternative IGCC designs may be available that give higher operating efficiencies and/or lower capital cost of plant.

One option is to avoid the high cost and efficiency losses of a syngas cooler by cleaning the syngas at high temperature. Porous ceramic (eg SiC) filters are currently the preferred option for particulate removal. Metal oxide systems may subsequently be used to remove hydrogen sulphide prior to gas combustion. [Prior hydrogen chloride and subsequent ammonia removal are also likely to be necessary.] The development of these gas clean-up systems requires considerable effort on the ways in which high temperature metal alloy and ceramic

Figure 4.7   Map Relating Consequent Need & Materials Technology

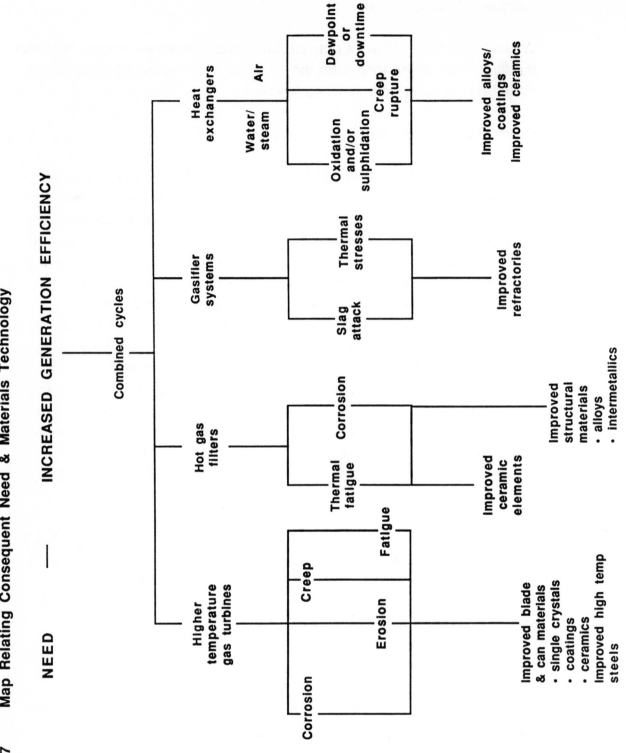

components may be used in combination.

These development requirements apply also to various hybrid options that combine fluidised bed combustion and coal gasification to increase operating efficiency by several percentage points. Gas from a partial gasifier is burned to boost the temperature of combustion gas from the fluidised bed before it enters the gas turbine in a so-called 'topping cycle'. British Coal have proposed one such cycle.

Development requirements for all of the above combined cycle plant options are summarised in Figure 4.7 which notes that the various gasification options can take advantage of developments of high temperature gas turbines discussed in Chapter 5.

Recent research and development has identified alternative, high efficiency gas turbine cycles which do not include a steam cycle. These feature an air humidifier after the compressor which lowers significantly the required compressor energy input. In baseload operation, a Humid Air Turbine (HAT) cycle when integrated with a coal gasification system, could be 10-15% lower in capital cost than a system with a steam generator.

For intermediate load operation, the Compressed Air Storage with Humidification (CASH) cycle is being developed for integration with coal gasification. These IGCASH plants are anticipated to save 15-20% in capital cost per peak kilowatt of capacity compared with a standard cycling pulverized coal plant with flue gas scrubbing. The combination of coal gasification and natural gas firing in a cycle called CASHING, compressed-air storage with humidification integrated with natural gas, could lower capital costs even further.

Longer Term Developments

Beyond the ten year plant timescale, there are a number of advanced plant options (eg involving indirectly-fired gas turbines) that require heat exchangers capable of withstanding metal temperatures approaching 1200°C. ODS alloys are being developed for this purpose. There is a requirement to achieve adequate high temperature strength and corrosion resistance and to avoid long term development of porosity, in these alloys.

Conversion of gas from a coal gasifier to electricity is possible in a fuel cell and a molten carbonate fuel cell, for example, can operate at gasifier exhaust temperatures. In principle, a gasifier-molten carbonate fuel cell

39

combination with heat recovery from gas leaving the cell could achieve efficiencies of 45-60%. This depends on developments in gas clean-up and fuel cell technology - see Figure 4.8.

Magneto hydrodynamic power plant has been considered for many years, but not developed commercially to date. Electricity is generated by passing very hot (hence ionised) coal combustion gas 'seeded' with alkali metal through a magnetic field. However, there are severe materials problems for a heat exchanger required to make use of the waste heat from the exhaust gases. In principle, efficiencies of 52-57% are possible.

Though additional capital cost is currently prohibitive, and options for safe disposal of CO_2 are uncertain, if greenhouse issues become more pressing there may be a future requirement to remove CO_2 from coal combustion gas. Some possible materials development requirements are mapped in Figure 4.9.

Figure 4.8 Map Relating Consequent Need & Materials Technology

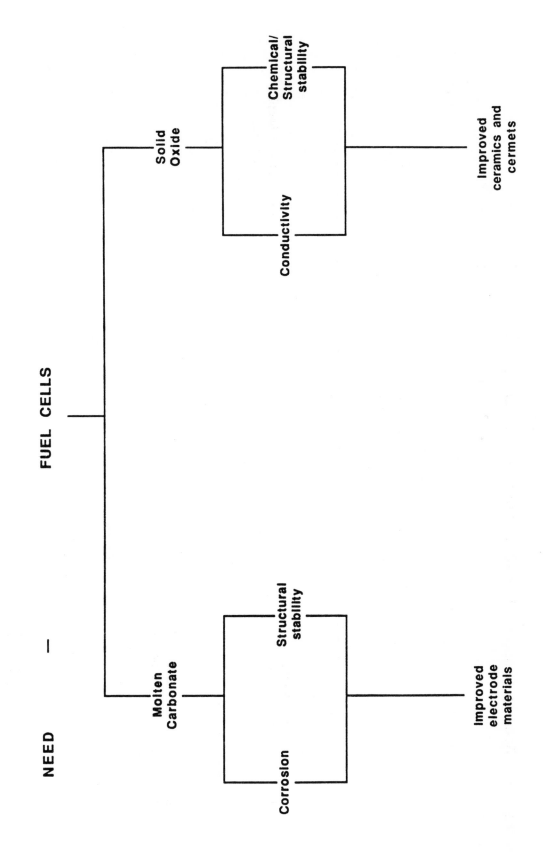

NEED

FUEL CELLS

Molten Carbonate

Corrosion

Structural stability

Improved electrode materials

Solid Oxide

Conductivity

Chemical/ Structural stability

Improved ceramics and cermets

Figure 4.9 Map Relating Consequent Need & Materials Technology

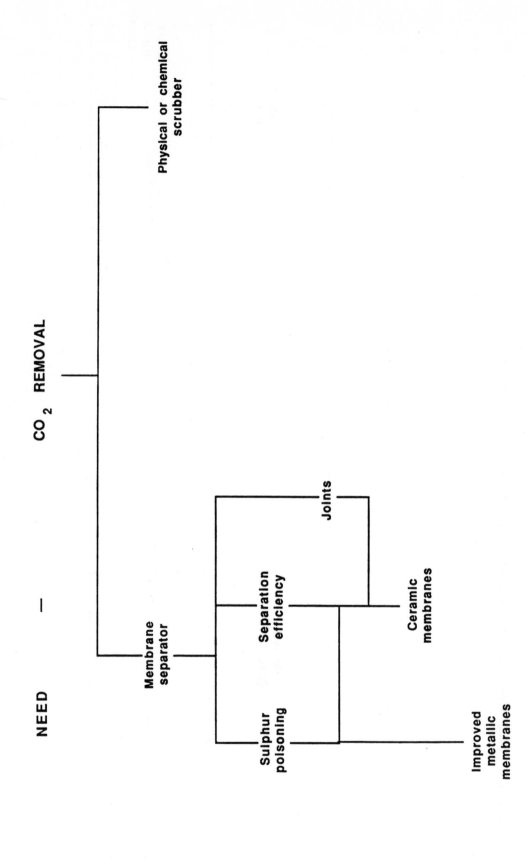

NEED — CO_2 REMOVAL

Physical or chemical scrubber

Membrane separator

Separation efficiency

Sulphur poisoning

Joints

Ceramic membranes

Improved metallic membranes

5. GAS

5.1 Introduction

The high efficiency and relatively low capital cost of combined cycle gas turbine plant, coupled with the availability of relatively cheap clean supplies of natural gas, have made CCGT the plant of choice for new capacity in a number of countries, including the UK. The challenge for the gas industry is to maintain secure low-cost supplies in the long term. For power plant manufacturers, there are pressures to continue to improve the performance of gas turbines. The latter developments will also improve the viability of coal gasification plant options.

5.2 UK Gas Supply Industry and Market Forces

In the early part of this century, the gas supplies in the UK were derived from coal via a simple gasification process. Following the discovery of natural gas (methane) fields in the North Sea, supplies were switched to natural gas, with accompanying modernisation of the distribution network. At this time the gas industry was in state ownership.

At first, the lifetime of reserves of natural gas available to the UK gas industry was uncertain. The industry therefore showed a keen interest in advanced coal gasification techniques, developing the British Gas/Lurgi slagging gasifier in collaboration with the German company, Lurgi, and taking this to the prototype stage at Westfield in Scotland. The potential for this gasifier to be used in coal-gasification combined cycle power generation schemes was recognised, the company being prepared to license the technology for that purpose. The threat of early exhaustion of gas reserves has, in fact, receded, and the latter 'clean coal' technology use is now considered to be the main 'raison d'etre' for the gasifier.

Over the last several decades, the traditional domestic markets for gas have been domestic cooking, domestic space heating and hot water heating, heating commercial and industrial premises, and industrial process heat, with progressive market expansion in the space heating sector, particularly following the oil price rises. To a much smaller extent, gas has been used in refrigeration and air conditioning systems. More than half the gas sold in the UK has been to these domestic markets.

The UK market for gas has been transformed as a result of privatisation of the gas supply industry, subsequent privatisation of the electricity supply industry, and deregulation of gas as a fuel for electricity generation.

This has caused the supply of gas for power generation in the UK to grow rapidly in the last five years, and rapid growth is expected to continue for at least another five years. Some of the major electricity generation and supply companies have become part-owners of gas reserves.

Cost considerations and environmental pressures are also favouring gas-fired CHP schemes for large commercial and industrial premises. Environmental pressures are leading to increased use of incineration as a disposal route for some difficult wastes, where gas-assisted firing can be an advantage. Though not taken up in the UK, some overseas coal-fired generating plant has been upgraded via supplementary gas burning to raise the steam temperature. These developments are providing a variety of additional opportunities for supply of gas.

The reserves-to-production ratio for natural gas is currently around 56 years (see Chapter 2), prompting a continuing search for new gas fields worldwide, both offshore and on land. The World Energy Council 1993 report (Energy for Tomorrow's World) states that, by 2020, both gas and oil reserves in Western Europe will have declined to a point at which only Norway is expected to have significant reserves of natural gas available. Future supplies of gas for the UK and Western Europe will, therefore, come in part from further afield; from Eastern Europe and from Northern Africa via long-distance pipeline networks, and from more remote supply areas in liquefied natural gas (LNG) tankers. Supplies may also come from the deeper water exploration areas on the western edge of the European continental shelf.

5.3 Technical Challenges of New Markets

The technical challenges for the gas industry, the market influences and the resulting materials technology implications are summarised in Tables 5.1 and 5.2.

Fuel Supply

The continued expansion of gas burning in the UK, coupled with the progressive exhaustion of the large gas fields found in the UK continental shelf in the 1960s and 1970s, has brought with it a need to maintain supplies by exploitation of smaller and more remote gas fields. To maintain the economic attractiveness of such developments, in an increasingly competitive energy market, it is necessary to reduce the through-life costs of all aspects of offshore operations, including drilling, the construction of production facilities and operation/maintenance activities. These requirements place greater emphasis on optimised

Table 5.1   Research Strategy Map for Gas Industry Materials Technology

Market Influence	Needs	Materials Technology Implications
(i) National & International Policies		
Environmental policies greenhouse gas concerns	High efficiency combustion New burner design Low NOx combustion Methane conversion to hydrogen Low gas leakage in transport	High temperature ceramic & metallic materials for combustors & heat exchangers Low cost catalysts and sensors Low cost catalysts and separation membranes Improved pipeline integrity
Conservation, recycling and incineration policies	Gas-assisted waste incineration Wood/gas combustion Coal bed gas development Land-fill gas, biomass	Corrosion resist. high temp metals & ceramics Low cost catalysts, separation membranes
Energy supply options	Low cost gas supply & distribution	Low cost structural & pipeline materials Low cost welding & assembly processes
Privatisation, regulation & deregulation	Safe, secure, reliable supply system	Fit-for-purpose inspection repair and maintenance systems
(ii) Gas Supply		
Smaller gas fields	Low cost drilling, low capital cost Offshore facilities	Optimised materials & fabrication selection (performance v installed cost) Subsea installations
	Low operating cost offshore facilities	Zero maintenance components
Less accessible gas fields	Low cost energy transport	Lower cost LNG transport Superconducting energy transport
Increasingly impure gas supplies	Low cost gas extraction, separation and treatment	Low cost separation membranes & catalysts
Alternative supplies	Mid-ocean exploration & production Deep mined methane hydrates	

Table 5.1 (cont'd)   Research Strategy Map for Gas Industry Materials Technology

Market Influence	Needs	Materials Technology Implications
(iii) Gas Transport, Storage & Distribution		
Increasing system complexity	Improved system control Accurate supply/demand monitoring Accurate metering & payment systems	Low cost energy meters
Rigid regulatory & financial regime	Low cost hook-up of new customers Plant life extension Zero leakage	Low cost pipeline materials & assembly tech. No-dig pipeline technology Low cost live inspection, repair & maint. Lower cost pipe lining & insertion methods Improved fogging & spraying, etc.
(iv) Gas Use		
Increasing competition in existing markets	Higher combustion efficiency Low NOx combustion Low capital/installed cost plant High reliability, zero maintenance plant	Lower cost high temp. ceramic & metallic mats. Low cost catalysts & sensors Optimised mats. sel. & production engineering Surface engineered components
Development of new gas uses	Market attractiveness for: Power gen. CHP Air conditioning Transport District heating & cooling Chemical feedstocks	Combined cycle gas turbine materials Long life engine & turbine component Fuel cell materials & assembly technology Low cost heat exchangers Catalysts, absorbents, membranes Lightweight CNG cylinders, absorbents Fuel cell materials & assembly technology Low cost insulated pipe systems Low cost catalysts & membranes

Table 5.2   Gas Industry - Performance Limitations to be Addressed by Materials R&D

Materials	Performance Limitations
High temperature metals and ceramics	Creep, corrosion/oxidation, thermal shock, toughness (ceramics) barrier coating integrity Low cost component manufacture, joining/assembly technology
Structural and pipeline steels	Corrosion performance in process fluids and external environments Inhibitor and coating performance; influence of temperature and flow turbulence. Low cost welding and joining.
Downhole and drilling materials	Wear-erosion-corrosion resistance, surface engineering of components. Performance assessment methods. Low cost component manufacture, joining and assembly technology, in-service degradation and assessment.
Structural polymers and reinforced plastics	Long term material properties, fluid permeability, design codes, performance testing. Surface performance, wear and in-service degradation, repair and refurbishment. NDE, quality and integrity. Low cost component manufacture.
Functional materials (fuel cells, catalysts, membranes, sensors, absorption materials, superconductors)	Nano-engineered bulk and surface properties. Manufacturing quality control, reliability and stability of functional performance. Assembly and joining technology. Low cost manufacture. In-service degradation (poisoning and fouling).

47

materials selection and fabrication processes, together with component reliability under conditions of reduced maintenance intervention.

Some natural gas supplies are impure, necessitating clean-up for their commercial exploitation. Membrane separation technology and catalytic conversion may offer advantages here.

The primary mode of gas transport is by high pressure pipelines. The existing pipeline infrastructure is increasing in age, and increased inspection and maintenance activity will be needed to maintain system integrity, including lower cost high integrity linings in some situations. Where new pipelines are required to exploit new smaller fields, untreated gas will require lower cost corrosion-resistant systems, preferably using coilable or re-usable pipelines. Higher operating pressures may also be required.

For long-distance transoceanic transport, LNG is currently the preferred option. Lower cost alternatives to the present LNG technology would be a particular advantage for remote gas field exploitation.

In the increasingly competitive gas supply market, more flexible gas storage systems are necessary. High pressure storage of natural gas could be enhanced by development of suitable absorbents.

A particular concern in transport, storage and distribution of gas is avoidance of leakage, particularly in the older parts of the low pressure distribution network. Apart from the obvious safety considerations, methane is a powerful greenhouse gas and a small percentage leakage could wipe out the greenhouse advantage of using gas as a fuel. Lower cost pipe lining and leakage control measures are required, especially those which can be undertaken without disrupting the gas flow or excavating large sections of pipe.

Power Generation

In the short and medium term, the power generation and CHP markets will take advantage of developments in gas turbines and engine technology, giving both higher efficiency and lower NO_x emissions. For other gas combustion applications, low NO_x is again a priority, with the possibility of catalytic combustion giving combustors tailored to special requirements. Ceramics and high temperature metals research for high temperature turbines and heat exchangers will benefit all aspects of gas turbine technology.

In the longer term, fuel cell technology is clearly a promising development at the commercial scale. Whether it might be scaled down to function economically at domestic size is unclear. The use of engines (such as the Stirling engine) for small scale CHP schemes is an alternative approach, if low-cost reliable low-maintenance systems can be developed.

5.4   GTs for Power Generation

Combined cycle gas turbine (CCGT) plant burning natural gas is clearly the favoured option for new large-scale generating capacity at this time. The worldwide demand for gas turbines has doubled in the last decade and around 25,000 MW of gas turbine power generation capacity is being installed every year. In the UK alone, the current level of 9,000 MW of natural gas fuelled combined cycle plant (which produces 15% of the baseload power requirements) is expected to increase to around 24,000 MW by the year 2000.

A simple cycle gas turbine power generation plant wastes heat in the exhaust gases and achieves an efficiency of 35%. [The new Rolls-Royce Trent Engine is expected to achieve 42% and will be the most efficient gas turbine in the world.] Higher efficiencies are obtained in CCGT plants, where the hot exhaust gases are used to raise steam in a Heat Recovery Steam Generator (HRSG) and the steam is then used to drive a steam turbine generator. The result is more power for the same fuel input, and efficiences of 55% are now being achieved (cf 47% ten years ago). This increased efficiency is achieved at the expense of higher capital costs for the additional plant, but CCGT plants are still considerably cheaper in capital cost than conventional nuclear, coal or oil fuelled plant.

Plant manufacturers are currently pursuing the development of large frame land based gas turbines and aeroderivative gas turbines in parallel (see Figure 5.1).

Improvements in the efficiency of gas turbines require higher values of combustion gas pressure and temperature at turbine entry, which in turn place onerous requirements on the creep, fatigue and corrosion performance of turbine blades and vanes. Nickel-based superalloys are the established materials for these components, and major improvements in efficiency to date have been achieved by improved designs, including blade cooling techniques. Further efficiency improvements require higher temperatures without increased blade cooling or, better still, with reduced cooling air flow. The required

Figure 5.1 Driving Forces for the Development of Gas Turbines

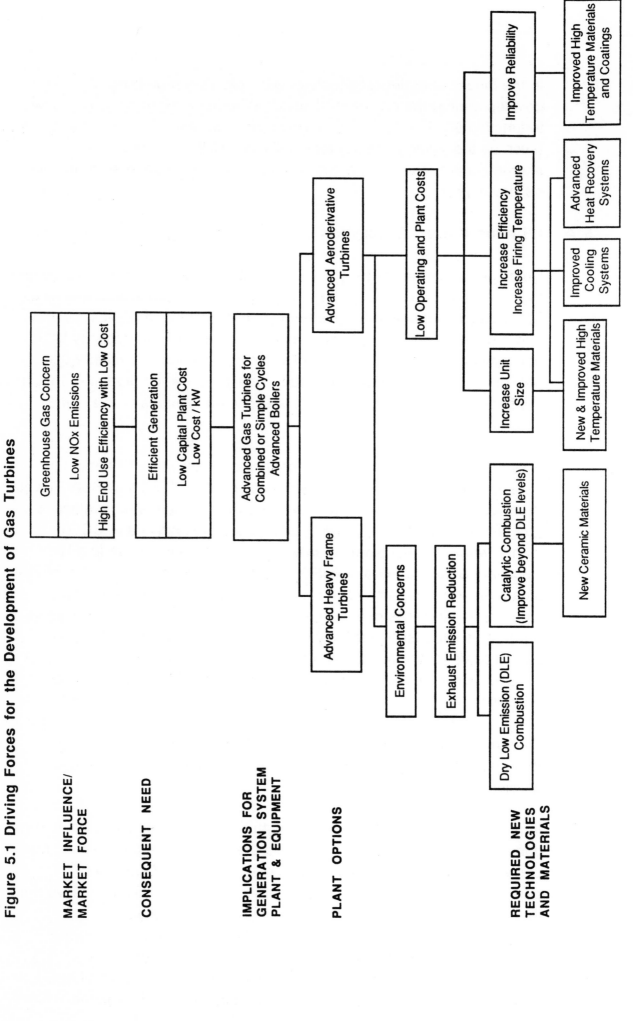

improved high temperature mechanical properties may be achieved by use of:

- Directionally-solidified alloys.

- Developments of single crystal blades.

- Development of powder metallurgy techniques to produce oxide dispersion strengthened blades.

Such developments are likely to give adequate mechanical performance, but still leave a shortfall in high-temperature corrosion resistance that requires additional development of:

- Corrosion resistant coatings, or

- Thermal barrier ceramic coatings.

All of these developments require considerable effort on creep, fatigue and corrosion resistance testing, integrity of castings, and fabrication techniques. Additionally, there is a need for improved understanding of, for example, fatigue of single crystal blades under the bi-axial stress conditions at their point of attachment.

Natural gas is a very clean fuel for use in gas turbines. As noted in Chapter 4, future power generation alternatives may be the cleaned-up gas from a gasifier, or the filtered hot gas from a pressurised fluidised bed. Other direct firing coal options are possible in the very long term. These fuels provide different corrosion characteristics that need to be covered in test programmes, and in some cases the requirement for increased erosion resistance. Many of the problems might be solved in the long term by development of ceramic turbine components, though development of more resistant protective coatings could be a cheaper option.

6. OIL

6.1 Introduction

The Chairman of the Executive Assembly of the World Energy Council, Gerhard Ott, has remarked that oil 'is, and will continue to be for the foreseeable future, the single most important source of world energy'. Though it has long since lost its position as one of the cheapest fuels in absolute terms, for some sectors, particularly transport, there are currently no viable alternatives to oil-derived products.

For large-scale power generation, there are often cheaper alternatives and oil is favoured only in special circumstances, discussed below. Large oil-fired power stations, built when the fuel was cheap, have been little used in the UK in the past decade. A new cheaper fuel, Orimulsion, may favour greater use of those assets. The oil industry may also need to develop a synergism with the power generation industry in order to develop clean technology options for disposing of high sulphur residuals.

6.2 External Influences

External influences on the oil industry are mapped in Figure 6.1.

The non-uniform geographical distribution of oil-reserves and the politics of the regions concerned have led to increases and instability in oil prices which have, in turn, altered the markets for oil. It is no longer a favoured fuel for space heating or large scale power generation.

The use of oil-derived products for transport is not entirely a captive market. Environmental concerns are promoting electrification of public transport and possible future use of electric or gas-powered vehicles. From the point of view of CO_2 emissions, oil sits between coal and gas in terms of CO_2 per unit of stored energy. However, the overall greenhouse balance needs to be investigated (from production through to end use) before judging particular applications.

Sulphur emissions from small engines that do not have SO_2 removal capability must be dealt with by the provision of low-sulphur fuel products. Hence there is a need for environmentally-acceptable disposal rates for high-sulphur residuals.

Figure 6.1

Map of market influences on oil industry

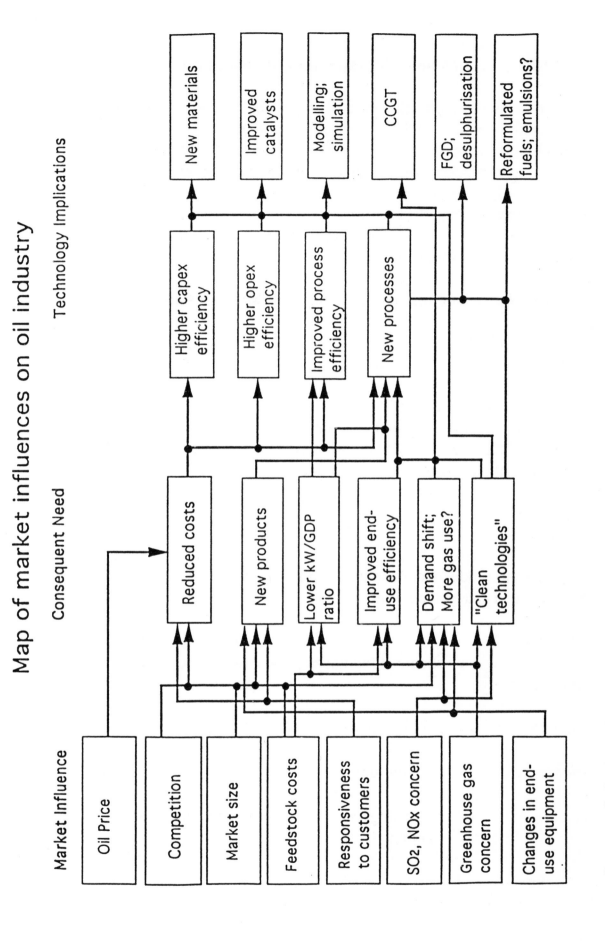

Source: D Barker, BP International Ltd

6.3 Extraction, Processing, Storage and Distribution

With a reserves-to-production ratio of only around 40 years (see Chapter 2), the search for new oil fields carries the same requirements as the search for new reservoirs of natural gas, both on land and offshore. Materials developments may reduce operation/maintenance costs and the frequency of component failure, and increase component life in a range of activities including:

construction of offshore platforms;
drilling;
provision of corrosion-resistant pipelines and wear-resistant pumps;
sea transport;
storage;
product distribution.

An oil refinery is a large chemical engineering plant, with the usual problems of materials integrity under high temperature hostile conditions. Developments in catalyst and membrane separation technology are important.

6.4 Utilisation

In outline, the products derived from crude oil via distillation (plus reforming and catalytic cracking) are:

refinery gas)
) $\rightarrow$ chemical feedstock $\rightarrow$ (industrial chemicals
naphtha   )   (fertilisers, plastics

(specialist chemicals

gasoline $\rightarrow$ motor vehicles

kerosene $\rightarrow$ aviation fuel

gasoil $\rightarrow$ diesel engines (land-based, road, rail, sea), heating

heavy fuel
oil $\rightarrow$ ships boilers, power stations

lubricating oil
paraffin wax

bitumen   →   road surfaces, waterproofing structures

Transport

Because of the high energy-to-weight ratio and light weight of storage requirements, oil-derived products are the major fuels of the transport industry.

The future materials requirements of the transport sector are the subject of a separate foresight study to be conducted by the Institute of Materials.

Power Generation

Some twenty years ago, world oil prices were low in relation to the prices of other fossil fuels, and this led to a major expansion of other markets, notably space heating and power generation. Since then there have been dramatic increases in the price of crude oil on three separate occasions, each time followed by some recovery towards lower levels. This substantially altered usage in the latter markets.

Large modern oil-fired power plant, built to take advantage of the original very low fuel prices, has been used very little in recent years in countries such as the UK with alternative generating capacity. It has continued to be used in countries with a shortage of alternative generating capacity, and has also been brought back on line when demand growth has been unexpectedly high (eg in Japan). Oil-fired plant is also used in States where oil is the indigenous energy source (eg Kuwait).

The convenience of small diesel generating sets for stand-alone applications has given them an expanding world market.

Large oil-fired power stations are subject to the same strictures as coal-fired power stations as far as SO_2, NO_x and particulate emissions are concerned. There has been a growing requirement to reduce emissions from diesel generators.

Many of the techniques and material developments required are common to the coal and gas industries already discussed, and they are not shown separately. However, they have often arisen first as requirements in the oil industry.

Heavy Residual Oils

With increasing pressures to restrict sulphur dioxide emissions, the oil industry has been required to produce lower sulphur products, leaving increasingly sulphur-rich heavy residuals for disposal. A favoured route for their disposal has been gasification to produce synthesis gas for methanol, ammonia or hydrogen production. With saturation of these markets, alternative disposal options are being sought. One possibility is as the fuel for a gasification combined cycle power generation scheme.

Orimulsion

Orimulsion is a relatively new fuel that can be used in oil-fired power stations and gasifiers. It is a water-based emulsion of bitumen obtained from large natural reserves in Venezuela. Competitive pricing could revive the fortunes of oil-fired plant that has been little used in the UK since oil price escalation, with corresponding proposals for Ince and Richborough plants by PowerGen and for Pembroke by National Power. Large scale long term use may require installation of FGD plant in order to meet company 'LCPD' limits or to satisfy HMIP local requirements, owing to the high sulphur content of the fuel.

7.   NUCLEAR POWER

7.1   Introduction

Figure 7.1 summarises the overall structure for nuclear power generation in terms of plant supply, its operation and maintenance during operation, the fuel supply and use cycle and the back-end decommissioning and waste management requirements. When assessing the economics of nuclear power generation, it is logical and convenient to break down costs into these elements.

A key distinguishing factor in nuclear electricity generation is that the heat source is nuclear fuel fission rather than combustion. This results in an intense irradiation environment in the reactor core which places particularly onerous requirements on the materials. The emissions from the fissioning of the fuel (ie fission products) are contained within the fuel cans and are part of the radwaste that has to be dealt with at the back end of the fuel cycle, as indicated in Figure 7.2.

The balance of forces affecting the position of nuclear power in the market place is rather different to the other electricity supply systems. These forces are outlined in the map in Figure 7.3 and their importance in drawing the consequent need through to material requirements is also outlined on the map.

Key influences for each of the four market forces on nuclear generation are identified, broadly showing:

• economics, driven by the need to reduce plant capital costs and achieve an improved return on capital investment through improved performance;

• environment, dominated by the back end waste management issues;

• strategic, where factors are generally positive for nuclear;

• socio-political, where public acceptability remains a key issue.

The remainder of the chapter amplifies on these main forces, particularly with regard to their influence on material requirements.

Figure 7.1   Nuclear Plant Cycle

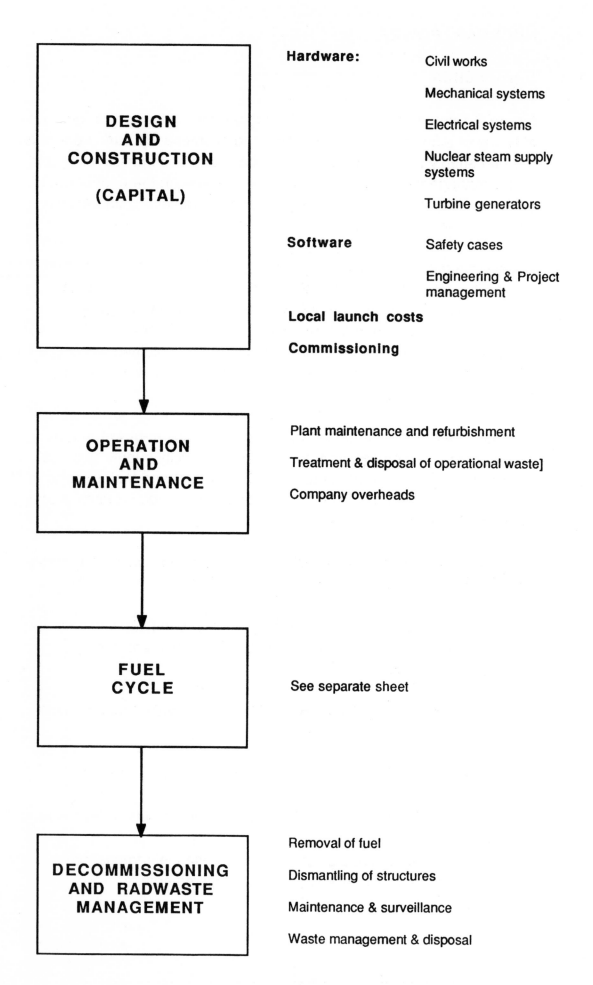

Hardware:	Civil works
	Mechanical systems
	Electrical systems
	Nuclear steam supply systems
	Turbine generators
Software	Safety cases
	Engineering & Project management

DESIGN AND CONSTRUCTION (CAPITAL)

Local launch costs

Commissioning

OPERATION AND MAINTENANCE

Plant maintenance and refurbishment

Treatment & disposal of operational waste]

Company overheads

FUEL CYCLE

See separate sheet

DECOMMISSIONING AND RADWASTE MANAGEMENT

Removal of fuel

Dismantling of structures

Maintenance & surveillance

Waste management & disposal

Figure 7.2   **The Nuclear Fuel Cycle**

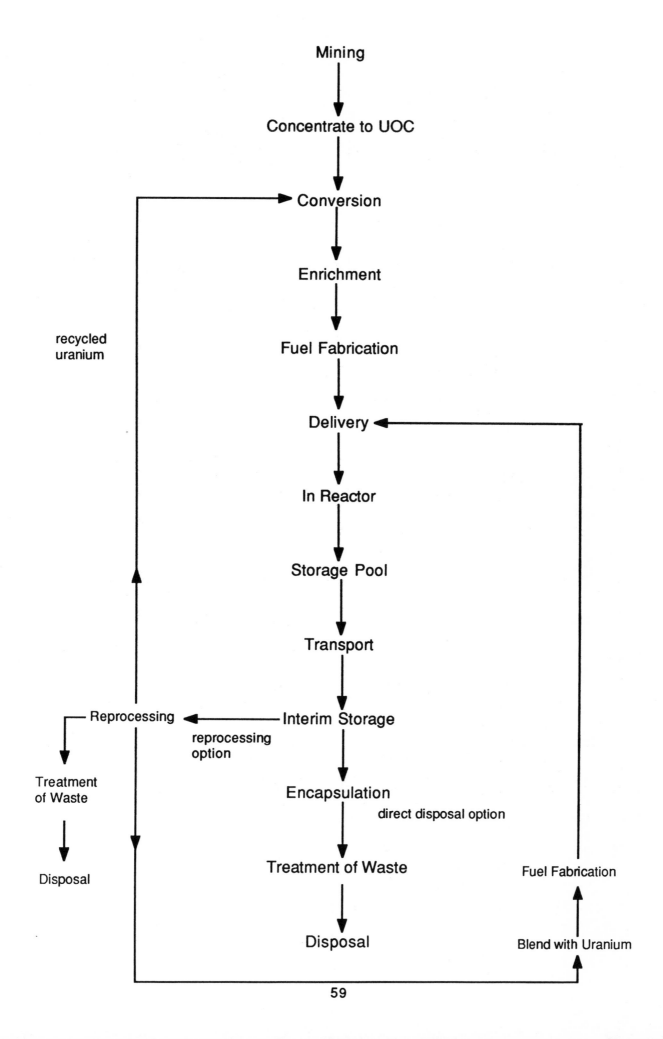

Figure 7.3 Map Relating Market Forces to Materials Requirements for Nuclear Power Industry

Column headings: MARKET FORCE | MARKET INFLUENCE | CONSEQUENT NEED | TECHNOLOGY IMPLICATIONS (a) | TECHNOLOGY IMPLICATIONS (b) | TECHNOLOGY IMPLICATIONS (c) | BARRIER OF PROGRESS | MATERIAL REQUIREMENTS

MARKET FORCE
- Economics
- Environmental
- Strategic
- Socio-political

MARKET INFLUENCE
- Price
- SO2, NOx concerns
- Greenhouse gas
- Waste cycle/fuel management
- Conservation of resources
- Diversity of supply
- Security of supply
- Non-fossil fuel obligation
- Export requirements
- Growth in electricity demand
- Pressures from personal transport
- Non-proliferation issues

CONSEQUENT NEED
- NUCLEAR

TECHNOLOGY IMPLICATIONS (a)
- Low Capital Cost
- Improved Plant Performance
- Waste Management

TECHNOLOGY IMPLICATIONS (b)
- Improved design & construction methods
- Fuel (1)
- Component Reliability
- Life Management

TECHNOLOGY IMPLICATIONS (c)
- Fabrication (4)
- Mechanical Life
- Core design (3)
- Enhanced Burn-Ups
- MOX (2)
- Steam Generator Unit (5)
- Primary Circuit Internals (6)
- Cables (7)
- Pumps/Valves (8)
- Monitoring/Inspection (9)
- Maintenance
- Repository performance
- Spent fuel storage
- Handling costs

BARRIER OF PROGRESS
- Powder Metallurgy Techniques
- Mechanical Interactions (Improved Design)
- Improved design, reflection
- Cladding Corrosion
- Hydriding
- Stress Corrosion
- Polymer Ageing
- Fibre Optic Ageing (11)
- Reduced Wear
- Improved Bearings
- Data Collection Hardware
- Monitoring Instrumentation
- Understanding of Degradation
- Coatings (12)
- Reduced Operator Dose
- Outage Management (13)
- Fuel clad corrosion (14)
- Volume reduction (15)
- Design of long term stores
- Spent fuel clad removal (16)
- Material activity levels (17)
- Surface decontamination (18)
- Decommissioning (19)

MATERIAL REQUIREMENTS
- Improved Processing Methods
- Enhanced Cladding Alloy
- Modified Coolant Chemistry
- Improved Material Understanding
- Monitoring/Operational Experience Inconel 690
- Improved Material Understanding
- Cobalt Free Materials
- Miniaturised Testing (10)
- Reduced Maintenance Requirements
- Improved Lifetime Performance
- Cobalt Free Pump Materials
- Automated Monitoring & Surveillance
- Enhanced cladding alloy
- Alternative structural materials & volume reduction techniques
- Improved understanding of fuel & cladding long-term integrity
- Chemical processes Alternative clad design/material eg Reduced use of cobalt
- Develop coating systems
- Design features to ease decommissioning

7.2 Structure, Market Influences and Technological Requirements of the UK Nuclear Industry

Market influences on the ESI have already been identified (Chapters 2 and 3) and have been grouped according to key market forces. In the maps in this section only those influences for which Nuclear is a reasonable option are shown.

The only realistic contender for future UK nuclear programmes in the timescales under consideration is the PWR design. Research and development within the UK is likely to be focused on PWR designs, and in considering technology implications and barriers to progress only these are considered.

Plant safety is an absolute requirement but it is recognised that modern PWRs meet very high standards and while improvements will always be sought, for example in the area of severe accident mitigation, plant economics remain a key issue. This relates to all aspects of the reactor plant cycle (see Figure 7.1):

- Reduced design and construction costs

- Reduced operating costs, achieved by lower fuel cycle costs and enhanced plant availability (extending plant life, fuel cycle operation and plant reliability)

- Reduced radioactive waste and ease of decommissioning.

The nuclear industry focus on achieving high standards is through the use of proven methods and materials: any material development is expected to be evolutionary and low risk.

Map Annotations

Fuel development

1. Main stream reactor development world-wide is in water-cooled reactor designs - PWR, BWR, PHWR. All use zirconium alloy fuel cladding and oxide fuel, usually UO_2 but with a developing use for mixed oxide $(U, Pu) O_2$. The main thrust of materials development is towards enhanced burn-ups, mainly seen as extending the time between re-fuelling which improves plant availability and overall economies rather than for simple fuel cycle cost reduction per se (although this is, of course, a factor to consider). [Burn up, in the

context of fuel element performance, is the amount of heat released or electricity generated in the fission of a given amount of fuel, expressed in GW days per tonne uranium.]

Principal design improvements to fuel assemblies are:

(i)   reduced mechanical interactions between fuel rods and other components

(ii)   reduced cladding corrosion

(iii)   improved resistance to hydriding.

Item (i) is almost entirely a mechanical design problem. Item (ii) is strongly influenced by coolant chemistry although some variant of cladding alloy with improved corrosion resistance may be developed; similarly for item (iii).

Because the fuel elements are replaceable and their performance is already very good, they are not life limiting or a major factor in improved economics, compared, for example, to improved reactor circuit material performance which could save very expensive and time consuming replacement operations.

2.   Use of mixed oxide (MOX) fuel to recycle energy. MOX is a key development in the shorter term, with investment in new plant planned.

3.   The core design can influence costs, with reducing the power density leading to greater safety/operating margins. There are lower fuel cycle costs linked to an increased number of in-core fuel batches for a given discharge burn-up. Incorporation of a radial reflector in the core design can reduce neutron leakage, leading to both increased core efficiency and reduction in the irradiation damage to the reactor vessel.

4.   Fuel fabrication is not seen as a major issue, although the process could benefit from technologies developed in the field of powder metallurgy.

Plant components reliability

5.   Steam generators have, to date, generally performed disappointingly. Tubes failures/degradation have dominated

recorded failure events. There has been considerable research activity, resulting in improvements in materials (eg, use of Inconel 690) and component design for the current and future generations of design. Monitoring is required to ensure long-term integrity of the new designs, which are largely unproven for extended periods of operation. It is anticipated that the plant operators will continue to seek greater reliability through materials and design improvements, stemming from improved understanding of the problem.

Two specific steam generator issues in the short term are:

a)   to confirm the reported immunity of Inconel 690 to primary water stress corrosion cracking for the higher pH primary chemistry regimes recently adopted

and

b)   to monitor the Westinghouse Model F (Sizewell B) anti-vibration bars to confirm that wear is not a problem at these locations.

6.   In addition to the steam generator issues, control rod drive mechanism tubes and other primary circuit components are susceptible to stress corrosion cracking associated with specific materials (such as Inconel 600). These remain major issues for PWRs, as they limit plant availability. Further materials research in the area is required in the short term.

7.   The lifetime of cables in existing plant requires assessment in the short term - this is linked to extending plant life.

8.   Improved bearings for pumps to enhance reliability and reduce maintenance requirements. Cobalt free pump materials can help to reduce operator dosage, helping to optimise maintenance and repair schedules. The objective is to avoid transportation of active materials in the fluid flowing through the system.

Life management

9.   Automated inspection/monitoring can be used to reduce outage time and hence increase plant availability. This looks to improved

data collection hardware, coupled with effective monitoring instrumentation (eg acoustic, vibration monitoring). There remains the need to process the data, and this leads to a fundamental requirement to understand material degradation processes, their manifestation and also critical regions of the structures.

10. Miniaturised testing seeks to maximise information from the minimum amount of material. The techniques are being developed on existing specimens, from broken pieces of larger scale tests. The objective is to be able to use scrapings, for instance, from the edge of the pressure vessel, to obtain plant performance data.

11. The increase in monitoring/control systems in plant is likely to lead to increased use of fibre optic cables. Research is required to assess their long term performance.

12. Studies are required to reduce the maintenance requirements of existing protective coatings (for example on concrete surfaces to avoid and/or help clean-up).

13. Planning and management of plant outages can be assisted by monitoring the surveillance information; effective planning has been recognised as a means of achieving cost reductions and maximising plant availability.

Waste Management/Ease of Decommissioning

14. Since the early days of nuclear power development, it has been appreciated that considerable attention needs to be given to waste management issues and particularly the requirements for minimising waste disposal costs. For instance, there are specific difficulties associated with corrosion of the UK magnox and material testing reactors' Magnox/Aluminium fuel cladding material in the fuel store and eventual repository environment; corrosion leads to gas formation, enhancing water movement and the transport of radionucleides. Modification of the cladding to give a low neutron cross-section material with reduced corrosion would enhance performance of the wasteform in the repository. Zircaloy behaves well in the fuel store environment, and is not expected to give problems during storage or eventual disposal.

15. In order to minimise costs associated with storage and disposal, consideration could be given to the development of materials and

techniques to assist volume reduction of active materials from the plant's structures and components.

16. Difficulties associated with removing the Zircaloy cladding for reprocessing of irradiated fuel adds significantly to its back-end cost. A mechanical shearing system is currently used. Alternative options, particularly for chemical removal of the cladding, could be examined.

17. Handling difficulties associated with certain materials can add to the overall plant costs. It is advantageous to avoid excessive use of materials with either long-lived nuclides or excessive gamma-emitting nuclides. For example, selecting materials to reduce cobalt pick-up/activity levels can reduce costs associated with ease of handling. This also impacts on operator exposure, and can also reduce plant maintenance costs (discussed above).

18. Surface coating systems for facilitating surface decontamination could be usefully developed. Examples are (strippable) paints without metal oxides in the filler. In epoxy based coatings, the filler also has greatest influence on the surface activities. An option is to use stainless steel as a surface material, and to decontaminate chemically or electrochemically.

19. Costs associated with decommissioning operations may be reduced by implementation of appropriate design features.

8. RENEWABLES

The term 'renewables' is now popularly used to describe both genuinely renewable sources of energy, such as biomass, and inexhaustible sources such as wind, wave power and solar energy. It has been further extended to embrace waste-to-energy schemes that have less adverse environmental impact than landfill or sea disposal.

Though renewable energy sources contribute some 20% of energy needs on a global basis, they amount to less than about 2% of UK generation at present. Additionally, most schemes for using renewable energy sources, apart from tidal barrage concepts, do not fall into the large power plant category (say > 300 MWe class) that is the main subject of this report. Therefore, no detailed discussion will be given here.

Renewable sources of energy have been critically reviewed recently by the Department of Trade and Industry (incorporating the former Department of Energy). To give some indication of the implications for required new and improved materials technology their assessment is summarised in Table 8.1.

Table 8.1 Renewable Sources of Energy and the Requirements for New and Improved Materials Technology[1]

Renewable Source	Rating[2]	Materials Requirements
SOLAR HEAT		
(i) Passive Solar Designs	Economically attractive	(i) Improved glass technologies, in particular: (a) Reflective and wear resistant coatings (metals, oxides and nitrides of copper, lead, silicon, titanium, tin, cobalt, iron, nickel, chromium, indium, silver and gold). (b) Coloured (body tinted) glass. (c) Variable transmission glass, double & triple glazing. (d) High performance thermally insulating smart glazing. (ii) Energy efficient timber framed houses.
(ii) Active Solar Heating	Long shot	(i) Lightweight, UV-resistant polymers and composites for solar panels. (ii) Extruded polymer radiators.
(iii) Photo Voltaic	Long shot	(i) Coatings to reduce reflective losses & weatherproofing. (ii) Cost Reduction. (iii) Larger single crystals of silicon for larger cells. (iv) Amorphous ribbons of semi-conducting silicon. (v) Other mass-produced semi-conductors, eg cadmium telluride. (vi) Thin layers of silicon applied to glass.

---

[1] Source - 'An assessment of Renewable Energy for the UK - ETSU ISBN 0-11-515348-9

[2] Assessment of above publication.

Table 8.1 cont'd

Renewable Source	Rating	Materials Requirements
BIO-FUELS		
(i) Solid Fuels from Municipal Waste	Economically attractive	(i) Systems which can handle wide range of corrosive residues - special requirements for compaction of waste, combustion systems, filtration, scrubbing of flue gases.
		(ii) Improved filtration materials - zeolites, activated charcoals.
(ii) Straw	Economically attractive	(i) Systems for compaction.
		(ii) Improved boiler materials.
(iii) Energy Forestry	Promising, but uncertain	(i) Evaluation of alternative crops for calorific value.
(iv) Other Fuels, eg poultry litter, sludges from paper making, lignin, etc	Promising, but uncertain	(i) Materials for improved fluidized bed technologies.
WIND POWER		
(i) On Land	Promising, but uncertain	(i) Improved fatigue and creep properties of blade materials.
		(ii) Reliable foam filled advanced composite structures.
		(iii) Low cost reliable wood or GRP composites.
		(iv) Improved NDT/condition monitoring techniques - self-monitoring materials.
(ii) Offshore	Long shot	(i) Improved corrosion resistance for materials of construction, especially welded structures.
		(ii) Improved marine fouling resistance.

Table 8.1 cont'd

Renewable Source		Rating	Materials Requirements	
WATER POWER				
(i)	Tidal	Promising, but uncertain	(i)	Improved corrosion protection systems for reinforced concrete structures.
			(ii)	Environmentally acceptable anti-fouling systems.
(ii)	Shoreline Wave Energy	Promising, but uncertain	(i)	Improved corrosion and marine fouling resistance of offshore structures.
(iii)	Offshore Wave Energy	Long shot		
GEOTHERMAL				
(i)	Hot dry rocks	Promising, but uncertain	(i)	Improved materials for hard rock drilling.
(ii)	Aquifiers	Long shot		

69

9. POWER PLANT MATERIALS RD&D IN THE UK: THE WAY FORWARD

9.1 Introduction

The preceding chapters have identified the demand for new large power generating plant, and the consequent requirement for materials research and development, by considering both UK and world requirements for new generating capacity. Figure 9.1 illustrates the interactions between R&D, the supply chain and the so-called facilitators.

If UK plant manufacturers and materials suppliers are to compete as effectively as possible in these markets, it is important that they have both:

- access to the necessary R&D, and

- opportunity to develop and demonstrate new products and new plant concepts, prior to their full commercialisation.

Each of these activities is increasingly international, but the position of UK-based manufacturers is clearly strengthened when there is a strong home base, or conversely weakened if such activities become concentrated in competitor countries.

9.2 The Changing Organisation of R&D

The re-structuring of the Electricity Supply Industry in the UK brought with it substantial changes in the way that power plant materials R&D is funded, and in the total level of funding. It also altered the basis on which any proposals for new plant demonstration need to be considered.

Prior to privatisation, the CEGB played a pivotal role in working together with the R&D centres in industry, government laboratories and universities in sponsoring power plant materials research. They played a major role themselves in leading materials R&D to underpin their power generation plants at their own laboratories (notably Leatherhead, Marchwood and Berkeley). But the other industry based R&D centres, eg plant manufacturers, British Steel and UKAEA also conducted major R&D programmes, frequently with financial support from the state-owned generating company. Together with CEGB, they stimulated and supported university research on materials.

Since privatisation, this infrastructure and support has declined for a complex range of reasons. A key element has been the view taken by the private and the newly privatised generators that their future requirements

Figure 9.1 Inter-relationships between the Main Supply Chain, Facilitators and R&D Centres

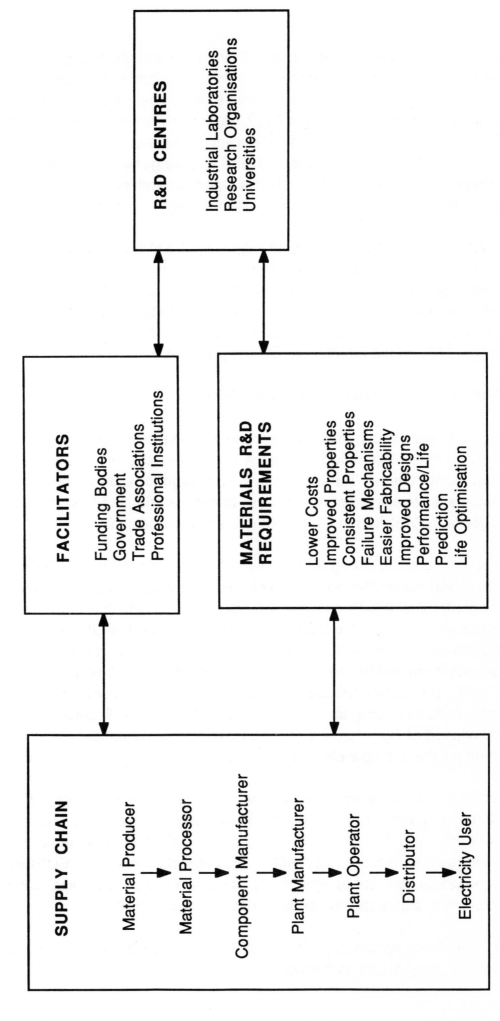

can be met by purchasing proven plant from the world's suppliers at the most competitive price. At the same time, UK plant manufacturers and materials suppliers, faced with a severe recession and competitive pressures, have reduced their R&D, particularly on longer term projects. A further factor has been a major decline in the support for nuclear power development. This is representative of a more general withdrawal by Government, notably DTI, of direct support for applied development programmes which it sees as the responsibility of industry.

Thus, the UK has been going through a period of transition with R&D partnerships tending to replace the old methods of funding. There is a clear need to continue and extend partnerships between the key players, including Government which is an important customer for R&D, particularly in the Universities, and which can influence the allocation of resources and the direction of EC programmes.

9.3 Partnerships between the Key Players

Such partnerships in materials technology R&D are increasingly international. This reflects the generic nature of the initial development of new plant material. The competitive advantage for a particular plant manufacturer arises not from access to the material, but from the way the material is exploited in the manufacturer's own designs.

For example, the strong ferritic steels (NF616, HCM12A and TB12M) described in Chapter 4 were developed by a project team comprising three steelmakers, three boilermakers and three countries' utilities, with representation from the USA, Japan, Denmark and the UK. The advantage of manufacturers' participation is a full understanding of the development and hence of the way in which it can be best exploited. For a utility, the same understanding allows a better assessment of rival manufacturers' designs. Utility participation is also indicative of a perception of likely future exploitation, in ultra-supercritical coal-fired plant for the example chosen.

Major collaborative activities are taking place within the European Community with CEC, and hence UK Government, support. Examples here are the COST and BRITE programmes which cover many of the materials requirements identified in previous chapters. The EUREKA organisation was set up specifically to foster R&D partnerships that will strengthen European competitiveness in a world market.

Turning more specifically to the roles of the key partners in the UK, these can be summarised as follows:

(a) Plant and Materials Suppliers

Plant manufacturers, together with the material suppliers, clearly have a pivotal role in power generation technology and materials development. This report has highlighted that the key drivers will be increasing competitive and environmental pressures on the generators, who will consequently require more efficient clean technology. The study has indicated that such advances will require both evolutionary and step changes in materials technology. An encouraging aspect of the analysis brought out in Chapters 4 - 8 is that there is no shortage of ideas on both short and long term technology requirements and the associated materials needs.

(b) Power Generators

The role of the power generators has clearly changed. The privatised companies can no longer act as a channel for state subsidy to manufacturers' R&D, and their duty to shareholders does not necessarily commit them to supporting UK-based manufacturing companies. This is certainly true for new entrants to the power generation market, where the major shareholder may even be an overseas plant manufacturing company. Nevertheless, it is in the interests of UK plant manufacturers to hold bi- and multi-lateral discussions with the generators, in order to establish a view on priorities and future requirements against which investment plans in technology developments and associated materials R&D can be assessed. Co-operation and partnership that can be established on the domestic front will serve to strengthen the ability of UK plant and material suppliers to compete effectively in overseas markets, as well as enabling the generators to choose best-value suppliers on an informed basis.

(c) Government

Recognising that Government will not revert to playing an interventionist role in directly sponsoring the development of generator technology, it nevertheless has an important role to play on a number of fronts.

First Government can help to facilitate the interaction and partnership between generators and manufacturers on the grounds of promoting improved industrial performance.

Second, and more specifically, Government can, and does, support European collaborative applied R&D programmes such as the EUREKA COST and BRITE programmes referred to earlier. It is essential that UK industry gains maximum benefits from such programmes.

Third, Government is a major provider of R&D funds through its support of the national science and engineering base. Following the White Paper 'Realising Our Potential', the intent to link this to wealth creation is more explicit. The current technology foresight exercise by OST encompasses Energy and Materials and this should provide a means of strengthening and focusing the Government-Industry partnership which could be of benefit to plant and materials suppliers through, for example, Research Council programmes that underpin the generator technology sector. Again, associated materials R&D must be an important component.

9.4 Demonstration Plants

Demonstration plants by their nature do not usually generate the return on investment required by a commercial operator. As the first plants of their type, their capital cost may be high. Plant availability may be low while initial design difficulties are being resolved, raising operational costs, and it may be some time before design efficiencies are achieved.

Nevertheless, new concepts must first be demonstrated. Around the world, various financial frameworks have been adopted in order to allow demonstrations to proceed. Generally these have involved direct or indirect Government support. In the US, for example, the Department of Energy has supported a substantial Clean Coal Technology Programme. Within Europe, the CEC is assisting the Puertollano IGCC project in Spain, in which the costs and risks are further shared by a number of participating utilities (including National Power from the UK). Another option, adopted for renewable technologies but not new coal-fired developments in the UK, is a subsidy as exemplified by the Non-Fossil Fuel Obligation.

Where there is a large state-owned Utility, one might expect that operator to bear the costs of demonstration. Indeed, in earlier years, the CEGB essentially provided the test bed for large generating sets and, together with British Coal, funded a substantial PFBC development programme at Grimethorpe.

Power generation in the UK is now in the hands of a number of smaller private generating companies, which cannot individually be expected to bear major plant development costs. As already remarked, the commercial pressures on these companies are to invest in proven least-cost technology.

If new power generation concepts are to be demonstrated in the UK, it will clearly be necessary to establish appropriate partnerships, together with an appropriate financial framework. This issue needs to be addressed at national level with Government involvement reflecting its responsibility for energy policy and national strategy, and its support for UK trade and industry.

10. FINDINGS AND RECOMMENDATIONS

10.1 Findings

(i)   The UK market for large power generating plant over the next ten years will continue to be dominated by the decisions of the major Utilities and new entrants to build combined cycle gas turbine plant. The popularity of natural gas as a fuel, on both cost and environmental grounds, has also created a substantial world market. However, there is a growing demand for coal-fired power plant in a number of developing countries (eg India, China) owing to their large indigenous reserves of this fuel. Market pressures are for more efficient and clean plant, and at the same time for lower capital cost per MW.

(ii)   There will be a major increase in electricity demand worldwide in the early part of the 21st century, led by the industrially developing nations. This will create major business opportunities for the main generating plant suppliers around the world. Over the longer term, environmental and fuel resource pressures emphasise the need for economic clean coal technology and safe, clean nuclear power. In the UK, a substantial amount of the currently employed ageing coal and nuclear powered generating capacity will need to be replaced in the first two decades of the next century. The type of plant chosen will be dictated by cost considerations, in the case of fossil fuels, and for nuclear power, by the outcome of the UK Government's Nuclear Review. These various markets are summarised in Figure 10.1.

(iii)   There is need for both evolutionary and some radical advances in materials technology to underpin the new power generation technologies that will be needed in the 21st century. Further efficiency improvements in gas turbines are required which implies higher inlet temperatures. They must also be made capable of withstanding the effects of impure coal-derived gases if certain advanced cycles are to be pursued. Advanced steam cycles and various coal gasification options require heat exchangers that retain their integrity at higher temperatures and, in the case of gasification, in particularly corrosive environments. For these applications, metallic materials are required with improved creep, corrosion, fatigue and wear resistance in the short and medium term, while developments in ODS alloys and ceramic materials will be needed for some of the advanced plant options early in the next century.

Figure 10.1

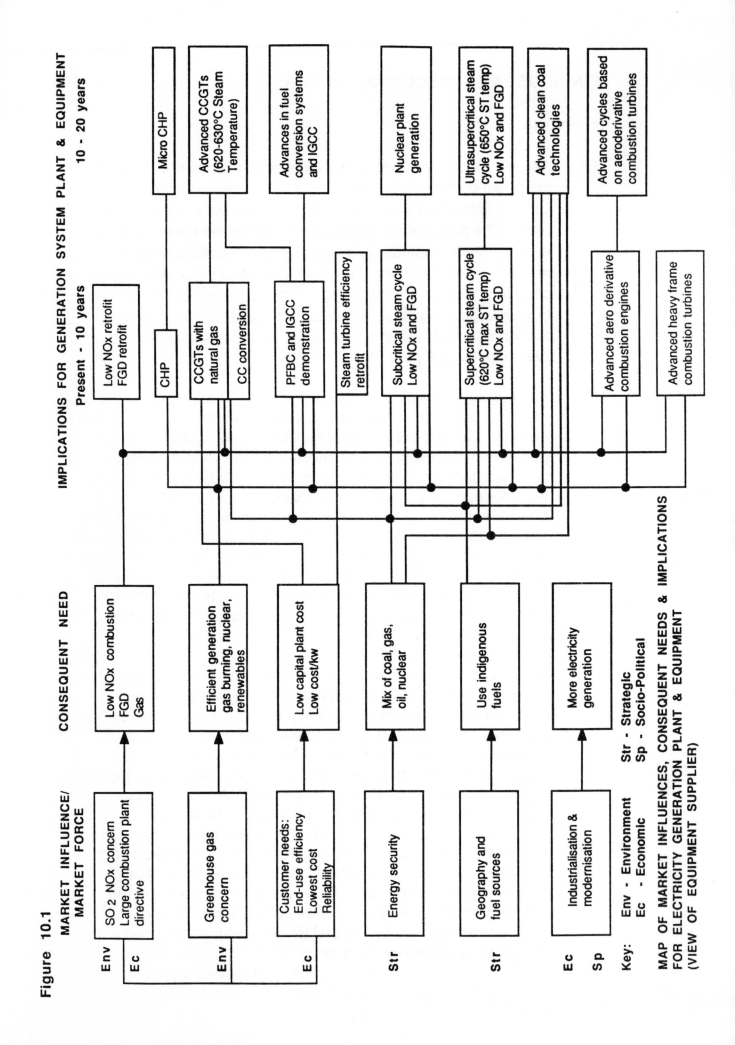

MARKET INFLUENCE/ MARKET FORCE

CONSEQUENT NEED

IMPLICATIONS FOR GENERATION SYSTEM PLANT & EQUIPMENT

Present - 10 years

10 - 20 years

Env SO2 NOx concern Large combustion plant directive

Ec

Env Greenhouse gas concern

Ec Customer needs: End-use efficiency Lowest cost Reliability

Str Energy security

Str Geography and fuel sources

Ec Industrialisation & modernisation

Sp

Low NOx combustion FGD Gas

Efficient generation gas burning, nuclear, renewables

Low capital plant cost Low cost/kw

Mix of coal, gas, oil, nuclear

Use indigenous fuels

More electricity generation

Low NOx retrofit FGD retrofit

CHP

CCGTs with natural gas

CC conversion

PFBC and IGCC demonstration

Steam turbine efficiency retrofit

Subcritical steam cycle Low NOx and FGD

Supercritical steam cycle (620°C max ST temp) Low NOx and FGD

Advanced aero derivative combustion engines

Advanced heavy frame combustion turbines

Micro CHP

Advanced CCGTs (620-630°C Steam Temperature)

Advances in fuel conversion systems and IGCC

Nuclear plant generation

Ultrasupercritical steam cycle (650°C ST temp) Low NOx and FGD

Advanced clean coal technologies

Advanced cycles based on aeroderivative combustion turbines

Key: Env - Environment Str - Strategic
Ec - Economic Sp - Socio-Political

MAP OF MARKET INFLUENCES, CONSEQUENT NEEDS & IMPLICATIONS FOR ELECTRICITY GENERATION PLANT & EQUIPMENT (VIEW OF EQUIPMENT SUPPLIER)

(iv)   For nuclear plant, reliability and operational efficiency will be improved by better understanding of stress corrosion and embrittlement mechanisms for core and steam circuit components and by use of cobalt-free materials to reduce dose to operators during maintenance. A range of materials integrity issues arise in the management of radio-active waste.

(v)   There has been a major decline in the last five years in UK funded R&D in support of new power generation systems. Various collaborative R&D programmes have been set up, or are under consideration, by the Utilities, plant manufacturers and material suppliers, in collaboration with independent research organisations and academia, some taking advantage of European Community funds. There is a need for a co-ordinated approach to UK power plant materials R&D that is designed to ensure that the overall programme is adequate, and that optimum use is made of available resources.

10.2 Recommendations

To Government, the power generators and the power plant and equipment makers and materials suppliers:

(i)   There is a need to continue and extend partnerships between these key players in order to ensure optimum use is made of national resources available to support power plant materials R&D. This co-operation must extend to ensuring we obtain maximum value from European and wider international collaborative programmes. More effective interactions stemming from a partnership approach are required to allow weaknesses in key areas to be identified and ensure that UK plant and materials suppliers are better prepared to compete in UK and world markets into the next century.

(ii)   Consideration should be given to means of funding demonstrations of new power plant technologies in the UK.

To Research Councils, Universities and Independent Research Organisations:

When considering funding priorities and decisions on longer term research strategies, these organisations should consider the materials needs of the UK power generation industry identified in this study. These include both evolutionary development of existing materials and a requirement for new materials for

77

advanced plant concepts.

To the Professional Engineering Institutions:

Making use of the results of this foresight exercise, The Institute of Materials could offer a forum for discussion of power plant materials strategy and, together with the relevant engineering institutions (eg I Mech E, IEE), assist in fostering collaboration between key interested parties on specific projects.